Studies in Systems, Decision and Control

Volume 331

The series "Studies in Systems, Decision and Control" (SSDC) covers both new developments and advances, as well as the state of the art, in the various areas of broadly perceived systems, decision making and control–quickly, up to date and with a high quality. The intent is to cover the theory, applications, and perspectives on the state of the art and future developments relevant to systems, decision making, control, complex processes and related areas, as embedded in the fields of engineering, computer science, physics, economics, social and life sciences, as well as the paradigms and methodologies behind them. The series contains monographs, textbooks, lecture notes and edited volumes in systems, decision making and control spanning the areas of Cyber-Physical Systems, Autonomous Systems, Sensor Networks, Control Systems, Energy Systems, Automotive Systems, Biological Systems, Vehicular Networking and Connected Vehicles, Aerospace Systems, Automation, Manufacturing, Smart Grids, Nonlinear Systems, Power Systems, Robotics, Social Systems, Economic Systems and other. Of particular value to both the contributors and the readership are the short publication timeframe and the world-wide distribution and exposure which enable both a wide and rapid dissemination of research output.

Indexed by SCOPUS, DBLP, WTI Frankfurt eG, zbMATH, SCImago.

All books published in the series are submitted for consideration in Web of Science.

More information about this series at http://www.springer.com/series/13304

Michal Mankowski · Mikhail Moshkov

Dynamic Programming Multi-Objective Combinatorial Optimization

Springer

Michal Mankowski
Computer, Electrical and Mathematical Sciences and Engineering Division
King Abdullah University of Science and Technology
Thuwal, Saudi Arabia

Mikhail Moshkov
Computer, Electrical and Mathematical Sciences and Engineering Division
King Abdullah University of Science and Technology
Thuwal, Saudi Arabia

ISSN 2198-4182 ISSN 2198-4190 (electronic)
Studies in Systems, Decision and Control
ISBN 978-3-030-63919-8 ISBN 978-3-030-63920-4 (eBook)
https://doi.org/10.1007/978-3-030-63920-4

This Springer imprint is published by the registered company Springer Nature Switzerland AG
The registered company address is: Gewerbestrasse 11, 6330 Cham, Switzerland

To our families

Preface

In this book, we consider extensions of dynamic programming for the study of multi-objective combinatorial optimization problems that allow us (i) to perform multi-stage optimization relative to a sequence of criteria, (ii) to describe the whole set of solutions or its essential part, (iii) to count the described solutions, and (iv) to construct the set of Pareto optimal points for a bi-criteria optimization problem.

We propose a fairly universal approach based on so-called circuits without repetitions in which each element is generated exactly one time. Such circuits represent the sets of elements under consideration (the sets of feasible solutions) and are used by counting, multi-stage, and bi-criteria optimization algorithms. For a given optimization problem, we should describe an appropriate circuit and cost functions. Then, we can use the designed algorithms for which we already have proofs of their correctness and ways to evaluate the required number of operations and the time.

We construct conventional (which work directly with elements) circuits without repetitions for matrix chain multiplication, global sequence alignment, optimal paths in directed graphs, binary search trees, convex polygon triangulation, line breaking (text justification), one-dimensional clustering, optimal bitonic tour, and segmented least squares. For these problems, we evaluate the number of operations and the time required by the optimization and counting algorithms and consider the results of computational experiments.

If we cannot find a conventional circuit without repetitions for a problem, we can either create algorithms for optimization and counting from scratch or can transform a circuit with repetitions into a so-called syntactical circuit, which is a circuit without repetitions that works not with elements but with formulas representing elements. We apply both approaches to the optimization of matchings in trees and apply the second approach to the 0/1 knapsack problem.

The results presented in this book can be useful for researchers in combinatorial optimization. The book can be used in courses for graduate students.

Thuwal, Saudi Arabia
August 2020

Michal Mankowski
Mikhail Moshkov

Acknowledgements

We are greatly indebted to King Abdullah University of Science and Technology for the immense support.

We are grateful to our coauthors in papers devoted to the design of multi-stage optimization algorithms for four classic combinatorial optimization problems and to former members of our group with whom we created various extensions of the dynamic programming approach: Hassan AbouEisha, Jewahir AbuBekr, Abdulaziz Alkhalid, Maram Alnafie, Fawaz Alsolami, Talha Amin, Mohammad Azad, Igor Chikalov, Shahid Hussain, Enas Odat, and Beata Zielosko.

We are thankful to Prof. Andrzej Skowron and to Alexander Knyazev for stimulating discussions.

We extend an expression of gratitude to Prof. Janusz Kacprzyk, to Dr. Thomas Ditzinger and to the Series Studies in Systems, Decision and Control staff at Springer for their support in making this book possible.

Thuwal, Saudi Arabia
August 2020

Michal Mankowski
Mikhail Moshkov

Contents

Chapter 1
Introduction

The dynamic programming has been studied intensively from its development by Richard Bellman in the 1950s [7] and applied to multiple domains such as control theory, economics, or in general, to optimization field. In this book, we consider dynamic programming in the realm of multi-objective (multi-criteria) combinatorial optimization. Our proposed algorithmic framework tackles two multi-objective optimization approaches, a priori and a posteriori optimization methods. In the a priori method, a combinatorial optimization problem is sequentially optimized with respect to several objectives ordered according to their significance. We refer to this approach as multi-stage optimization. It can be considered as a special case of lexicographic optimization. Contrarily, the a posteriori method provides an opportunity to revoke an order of objectives resulting in simultaneous optimization relative to two criteria. The aim is to find the set of Pareto optimal points related to the Pareto optimal solutions, which cannot be improved without worsening at least one objective. We refer to this approach as bi-criteria optimization.

The conventional dynamic programming algorithms for solving combinatorial optimization problems include a structure of subproblems of the initial problem, a way to construct a solution to a subproblem from solutions to smaller subproblems, and solutions to the smallest subproblems. Such algorithms return only one solution.

In this book, we consider extensions of the dynamic programming approach for the study of multi-objective combinatorial optimization problems that allow us (i) to perform multi-stage optimization relative to a sequence of criteria, (ii) to describe the whole set of solutions or its essential part, (iii) to count the described solutions, and (iv) to construct the set of Pareto optimal points for a bi-criteria optimization problem.

Earlier, we designed multi-stage and bi-criteria optimization algorithms for decision trees, decision rules, and element partition trees that are used for solving partial differential equations—see book [2]. In the same book, we considered multi-stage optimization algorithms for four classic combinatorial optimization problems. Each problem was studied separately with its own problem-specific algorithms and proofs.

M. Mankowski and M. Moshkov, *Dynamic Programming Multi-Objective Combinatorial Optimization*, Studies in Systems, Decision and Control 331,
https://doi.org/10.1007/978-3-030-63920-4_1

In this book, we consider a fairly universal approach to work with combinatorial optimization problems based on the use of so-called circuits without repetitions in which each element is generated exactly one time. These circuits were introduced in [38] (see also [39]). Such circuits represent the sets of elements under consideration (the sets of feasible solutions) and are used by counting, multi-stage optimization, and bi-criteria optimization algorithms. For a given combinatorial optimization problem, we should describe an appropriate circuit and cost functions. Then, the optimization and counting algorithms can be applied. For these algorithms, our framework provides the proofs of correctness and ways to evaluate the number of required operations and the time complexity based on the description of the circuit and cost functions.

In the first part of the book, we study circuits without repetitions, cost functions for these circuits, and algorithms for counting elements represented by a circuit and for optimization of these elements. Besides the conventional circuits that work directly with elements, we also consider so-called syntactical circuits, which work with formulas representing elements.

In the second part of the book, we construct conventional circuits without repetitions for nine known combinatorial optimization problems, evaluate the number of operations and the time required by counting and optimization algorithms, and consider the results of experiments.

We could not find a conventional circuit without repetitions that describes the whole set of elements under consideration (the set of feasible solutions) for each combinatorial optimization problem. One such problem is the optimization of matchings in a tree. In the third part of this book, we design algorithms for counting the number of matchings, and for multi-stage and bi-criteria optimization of matchings, analyze these algorithms, and consider the results of computational experiments.

In the fourth part of the book, we consider another approach to work with problems for which we could not find conventional circuits without repetitions. In this case, we can transform a conventional circuit with repetitions into a corresponding syntactical circuit. This syntactical circuit is a circuit without repetitions. We study syntactical circuits for two problems: optimization of matchings in trees and 0/1 knapsack problem.

1.1 Circuits Without Repetitions and Combinatorial Optimization Problems

In describing a single-criterion optimization problem, we usually define a set of elements (a set of feasible solutions) and a cost function on this set. Here, we approach this description task by using the notion of a circuit [38, 39]. Our circuit builds a set of elements from one-element sets attached to input nodes. It uses the operation of union of sets attached to unifying nodes and functional operations attached to functional nodes. It is, in some sense, similar to the structure of subproblems in a traditional dynamic programming algorithm. We will consider these circuits that work directly with elements as conventional circuits.

There is some similarity between the circuits considered in this book and combinatorial circuits that compute Boolean functions [37, 52, 58]. However, the circuits studied in this book are closer to $\{\cup, \cap, ^-, +, \times\}$-circuits computing (representing) sets of natural numbers [41] or integers [55].

We assign a set of elements $S(v)$ to each node v of a circuit S and define a cost function Ψ on the union of these sets. The function Ψ corresponds to each element δ a real number $\Psi(\delta)$—the cost of δ. One of the nodes v_* of the circuit S is considered as its output. The set of elements $S(v_*)$ (considered as the set of feasible solutions) and the cost function Ψ form the optimization problem in which we minimize the cost of elements from $S(v_*)$ relative to the cost function Ψ. Elements from $S(v_*)$ with the minimum cost are called optimal in $S(v_*)$ relative to Ψ.

We mainly study so-called circuits without repetitions in which sets of elements corresponding to input and functional nodes are pairwise disjoint and in which functions attached to functional nodes are injective. The use of such circuits guarantees consistent cost for all elements (the same elements have the same cost) and simplifies the counting of elements described by the circuit. We study for our circuits only the increasing and strictly increasing cost functions.

We consider an optimization procedure introduced in [39] (algorithm $\mathscr{A}_1$), which transforms the circuit S without repetitions into its subcircuit S^Ψ obtained from S by the removal of some edges entering unifying nodes. This subcircuit is also a circuit without repetitions. For strictly increasing cost functions, the set of elements $S^\Psi(v_*)$ assigned to the output v_* of the subcircuit S^Ψ coincides with the set of elements from $S(v_*)$, which are optimal in $S(v_*)$ relative to Ψ. With increasing cost functions, the set of elements $S^\Psi(v_*)$ is a nonempty subset of the set of elements from $S(v_*)$, which are optimal in $S(v_*)$ relative to Ψ.

The algorithm $\mathscr{A}_1$ can be used for multi-stage optimization of elements from $S(v_*)$ relative to a sequence of increasing or strictly increasing cost functions $\Psi_1, \ldots, \Psi_t$. We also consider an algorithm $\mathscr{A}_2$ introduced in [39] for counting the cardinality of the set $S(v_*)$ and the sets $S^{\Psi_1,\ldots,\Psi_i}(v_*)$ obtained after the ith step of optimization, $i = 1, \ldots, t$.

We pay special attention to bi-criteria optimization problems relative to two increasing cost functions. Our algorithm $\mathscr{A}_4$ introduced in [38] allows us to construct the set of Pareto optimal points for such a problem. We can use this algorithm to recognize the existence of an element from $S(v_*)$ that is optimal relative to two increasing cost functions Ψ_1 and Ψ_2 at the same time. We call such elements totally optimal relative to Ψ_1 and Ψ_2. A totally optimal element exists if and only if there is only one Pareto optimal point for the bi-criteria optimization problem relative to Ψ_1 and Ψ_2.

The elements from the sets attached to the input nodes of a circuit and the functions attached to the functional nodes of this circuit can be complicated. We show that the algorithms $\mathscr{A}_1$, $\mathscr{A}_2$, and $\mathscr{A}_4$ can work not only with circuits but also with schemes of circuits obtained from circuits by the removal of one-element sets attached to input nodes, functions attached to functional nodes, and operations of union attached to unifying nodes.

We tested our proposed approach on the following nine combinatorial optimization problems: matrix chain multiplication [26], global sequence alignment [44], optimal paths in directed graphs [20, 61], binary search trees [35], convex polygon triangulation [25, 34], line breaking (text justification) [36], one-dimensional clustering [60], optimal bitonic tour [16], and segmented least squares [33]. Multi-stage optimization algorithms for the first four problems were developed earlier [3, 4, 12, 15], although each problem was studied separately with its own problem-specific algorithms and proofs.

Each of these problems has inherent dimensions—usually one or two natural numbers that describe the structure of problem instance. The problem instances include polynomial in dimensions number of numeric parameters. For each of the considered problems, each problem dimension is at most the number of numeric parameters in problem representation plus one.

For the problem of matrix chain multiplication, the dimension is the number of matrices n and the numeric parameters are $n+1$ natural numbers $m_0, m_1, \ldots, m_n$ that define sizes $m_0 \times m_1, m_1 \times m_2, \ldots, m_{n-1} \times m_n$ of these matrices. For the global sequence alignment problem, the dimensions are the lengths m and n of sequences. An instance of this problem has $m+n+mn$ numeric parameters that describe the costs of alignment of symbols and alignment of symbols and gaps.

For each of the considered problems, we describe the corresponding conventional circuit without repetitions, evaluate the time complexity of this circuit construction, discuss at least two increasing or strictly increasing cost functions, evaluate the number of operations (there are operations of the kind $x+y$, $x-y$, $x \cdot y$, x/y, $\min(x, y)$, $\max(x, y)$, $\sqrt{x}$, $\lceil x \rceil$, $\lfloor x \rfloor$, $|x|$) and the time required by the algorithms $\mathscr{A}_1$, $\mathscr{A}_2$, and $\mathscr{A}_4$, discuss an example, and provide results of experiments.

For each problem and each of the considered cost functions, the algorithms $\mathscr{A}_1$ and $\mathscr{A}_2$ require polynomial number of operations depending on the problem dimensions. In many cases, the algorithm $\mathscr{A}_4$ requires pseudo-polynomial number of operations (a polynomial number of operations depending on the problem dimensions and some numeric parameters of the problem). For example, for the matrix chain multiplication problem, we use the numeric parameter $\max\{m_0, m_1, \ldots, m_n\}$. For optimal paths in directed graphs, line breaking, one-dimensional clustering, convex polygon triangulation, optimal bitonic tour, and segmented least squares for some pairs of cost functions, the algorithm $\mathscr{A}_4$ requires polynomial number of operations.

Evaluating time complexity, we study two models of computation: the software model and the integer model. The software model is closely related to the conventional software when each of the considered operations requires a small constant number of clock cycles and we can use both variables for integers and variables for floating-point numbers. This model assumes that we are working with finite sets of numbers, for example, with single-precision (32-bit) floating-point numbers and 32-bit signed integers. For the software model, from the existence of polynomial or pseudo-polynomial bounds on the number of operations, it follows that the considered algorithms have polynomial or pseudo-polynomial time complexities, respectively.

The obtained polynomial bounds depend on the problem dimensions. The obtained pseudo-polynomial bounds depend on the problem dimensions and some numeric parameters of the problem.

The integer model assumes that all inputs are integers of arbitrary size, and each operation transforms a tuple of integers into an integer. In this case, the time complexity of an operation depends on the maximum number of bits t in the binary representation of integers at the operation inputs. For example, addition, subtraction, and comparison require $O(t)$ and the multiplication requires $O(t \cdot \log t \cdot \log\log t)$ time [51]. For the integer model for each of the considered cases when the problem instance is represented by integer parameters and we use so-called integer cost functions, we can transform polynomial bounds on the number of operations into polynomial bounds on the time complexity of algorithms depending on the length of the problem representation. We also can transform pseudo-polynomial bounds on the number of operations into pseudo-polynomial bounds on the time complexity of algorithms (polynomial depending on the length of the problem representation and some numeric parameters of the problem). This way to obtain bounds on the time complexity of algorithms (evaluation of the number of operations and the size of used integers) is close to the notion of strongly polynomial time [27].

1.2 Beyond Conventional Circuits Without Repetitions

We cannot construct a conventional circuit without repetitions for each combinatorial optimization problem. For some problems, a solution to a subproblem may be a solution to the initial problem. A matching in a subtree of a tree is a matching in the tree. It is also true in the general case for decision trees and decision rules. A decision tree can be appropriate both for a decision table and its proper subtable. The same situation is with decision rules. In this case, we should find a way to describe all elements under consideration (the set of feasible solutions). Usually, it is a labeled directed acyclic graph (DAG), which nodes are related to subproblems of the considered problem [2]. After that, we should design and analyze counting, multi-stage optimization, and bi-criteria optimization algorithms based on this DAG.

We apply this approach to the problem of optimization of matchings in a tree relative to weight functions that assign positive integers to edges of the tree. We did not find a conventional circuit without repetitions that represents the set of all matchings for an arbitrary tree. Instead of a circuit, we construct a labeled forest to describe the set of matchings. We use this forest in multi-stage and bi-criteria optimization of matchings and in counting matchings.

There is also another way to work with the case when we can construct only a conventional circuit S with repetitions for a given problem. We transform this circuit into a syntactical circuit $\hat{S}$, which works with formulas representing elements. In these formulas, we use element symbols instead of elements and functional symbols instead of functions. The used symbols are pairwise different. The circuit $\hat{S}$ is a circuit without repetitions. We can define cost functions for $\hat{S}$. However, we should be sure that different formulas from $\hat{S}(v_*)$ representing the same element have the same cost. We can apply to this circuit the algorithms $\mathscr{A}_1$, $\mathscr{A}_2$, and $\mathscr{A}_4$. Note that the algorithm $\mathscr{A}_2$ will return not the number of elements but the number of formulas (the same element can have two or more formulas representing it).

We use this way to study optimization of matchings in trees and 0/1 knapsack problem. For optimization of matchings in trees, the algorithms $\mathscr{A}_1$ and $\mathscr{A}_2$ require polynomial number of operations, and the algorithm $\mathscr{A}_4$ requires pseudo-polynomial number of operations. For 0/1 knapsack problem, the algorithms $\mathscr{A}_1$, $\mathscr{A}_2$, and $\mathscr{A}_4$ require pseudo-polynomial number of operations. As for the conventional circuits without repetitions, we can transform the obtained bounds on the number of operations into bounds on the time complexity of algorithms in the frameworks of the two models of computation. The only exception is the time complexity of the algorithm $\mathscr{A}_2$ in the framework of the integer model. We have no tools to evaluate the number of formulas described by a syntactical circuit. As a result, we cannot evaluate the size of numbers during the work of the algorithm $\mathscr{A}_2$.

1.3 Comparison with Other Investigations

Our results are related to the fast-growing study of multi-objective (multi-criteria) optimization [18]. Multi-stage optimization belongs to the group of a priori methods for solving multi-criteria optimization problems, where the order of criteria is predefined. Multi-stage optimization can be considered as a special case of lexicographic optimization. The essential feature of the latter approach is that the optimality for the kth criterion is achieved once optimality for the first $k-1$ criteria is accomplished. Bi-criteria optimization belongs to the group of a posteriori methods related to the construction of Pareto optimal points and Pareto optimal solutions.

In this section, we compare our study (when it is possible) with other investigations. Sometimes, bounds on the time complexity obtained in this book are greater than the bounds obtained in special investigations of concrete problems. It is understandable since our approach is fairly universal.

Earlier, we worked on multi-stage optimization of decision trees [42, 43] and designed algorithms for multi-stage optimization of decision rules, decision rule systems, and inhibitory trees [2, 5]. Later, we also looked at the relationships between different parameters of decision trees including the construction of the set of Pareto optimal points [11, 13, 31]. Our bi-criteria optimization algorithms for the decision trees, rules, and rule systems can be found in [2, 14]. We also studied multi-stage and bi-criteria optimization of element partition trees that are used in finite element methods for solving partial differential equations [2]. To the best of our knowledge, no other results related to multi-stage and bi-criteria dynamic programming optimization of decision trees, rules and rule systems, and element partition trees have been published.

The situation with classic combinatorial optimization problems is different. We should mention here an Algebraic Dynamic Programming (ADP) approach, which was created by Robert Giegerich and his colleagues mainly for the bioinformatics applications [22, 23]. In this approach, the problem decomposition is described by a tree grammar and the optimization criterion is given by an evaluation algebra satisfying Bellman's principle. Initially, ADP included optimization relative to one criterion. Later, lexicographic optimization was added to ADP [24] (see also [54]). The construction of the set of Pareto optimal points for bi-criteria optimization problems was incorporated into ADP in [21, 49, 50]. Besides clear differences, ADP and our approach have some common features.

Looking beyond ADP and conventional mathematical programming methods, it is hard to find general frameworks designed for combinatorial lexicographic optimization. Volgenant [57] proposed a fairly broad approach to lexicographic optimization of certain graph theory problems such as shortest path, minimum spanning tree, and linear assignment. For multiple combinatorial optimization problems, problem-specific lexicographic optimization algorithms have been developed, including problems such as Steiner tree [59], network flow [8, 9], location [28, 45], or cutting stock [53].

Earlier, we proposed multi-stage optimization algorithms for the following four combinatorial optimization problems: matrix chain multiplication, global sequence alignment, optimal paths in directed graphs, and binary search trees [3, 4, 12, 15]. Later, in [39], we designed multi-stage optimization algorithms for convex polygon triangulation, one-dimensional clustering, and line breaking (text justification) problems. We were not able to find other publications related to multi-stage optimization of ten out of 11 combinatorial optimization problems considered in this book. The only exception is the problem of optimal paths in directed graphs. The bound on the time complexity $O(n^2)$ obtained in [57] is better than the bound $O(mn^2)$ considered in Sect. 7.2, where n is the number of nodes and m is the number of edges in the graph. Both bounds are obtained in the framework of the software model of computation.

In [38], we proposed bi-criteria optimization algorithms for the following three combinatorial optimization problems: matrix chain multiplication, binary search trees, and optimal paths in directed graphs. We did not find other publications related to the algorithms for multi-criteria optimization of matrix chain multiplication, binary search trees, convex polygon triangulations, one-dimensional clusterings, bitonic tours, segmented least squares, and matchings in trees that construct the whole set of Pareto optimal points. However, such publications exist for the optimization of sequence alignments, paths in directed graphs, line breakings, and solutions to 0/1 knapsack problem.

For bi-criteria optimization of sequence alignments, the polynomial time dynamic programming algorithm to generate all Pareto optimal points was introduced in [48] and improved in [1]. We can compare only results obtained in the framework of the software model of computation for the case when one of the cost functions is the number of indels (matches of letters and gaps). The bound $O(mn(m+n))$ obtained in [1] is better than the bound $O(mn(m+n)\log(m+n))$ from Sect. 6.2, where m and n are the lengths of the considered sequences.

The paper [29] contains complexity analysis for ten bi-objective optimization problems related to the paths in directed graphs. For each problem, the aim is to construct a minimum set of paths for which pairs of values of the considered cost functions form the set of Pareto optimal points. These problems include, in some sense, all problems considered in Chap. 7. We obtained similar results: the algorithm $\mathscr{A}_4$ is polynomial in the cases when the paper [29] shows the existence of polynomial algorithms, and it is pseudo-polynomial when the paper [29] shows the existence of pseudo-polynomial algorithms only since the cardinality of the set of Pareto optimal points can be exponential in the length of the input. However, the bounds on the number of operations obtained in [29] are less than the bounds obtained in Sect. 7.2 (see a discussion of bounds in this section). A comparison of different approaches to solve bi-objective shortest path problems can be found in [46].

Holkner [30] studied multi-objective line breaking problems and a dynamic programming algorithm that constructs the sets of Pareto optimal points for these problems. The time complexity of the algorithm was not reported.

The methods presented in [32] were early attempts to apply dynamic programming to the multi-objective knapsack problem. Certain improvements in the dynamic programming approach were achieved by using dominance relations to discard some partial solutions [6]. The dynamic programming was also used in [17, 19, 47]. Captivo et al. [10] transformed the knapsack problem into a bi-objective shortest path problem that was solved using a labeling algorithm. Another method commonly used is a branch and bound [40], sometimes incorporated into the second phase of the two-phases method to solve the bi-objective knapsack problem [56].

1.4 Book Contents

This book includes an introduction and four parts consisting of 18 chapters.

1.4.1 Part I. Circuits Without Repetitions. Main Tools

In the first part, we consider notions and tools that are common for all chapters in Parts II and IV. The first part contains three chapters.

In Chap. 2, we discuss four basic notions: the notion of a circuit, the notion of a circuit without repetitions, the notion of a cost function for a circuit, and the notion of a syntactical circuit corresponding to a given circuit. We also consider four topics related to the algorithms over circuits: schemes of circuits, adaptive cost functions, bounds on the number of operations, and bounds on the time complexity.

In Chap. 3, we propose an algorithm $\mathscr{A}_1$, which can be used for multi-stage optimization of elements relative to a sequence of cost functions, and an algorithm $\mathscr{A}_2$ for finding the cardinalities of the sets of elements described by the initial circuit and its subcircuits obtained during the process of multi-stage optimization.

In Chap. 4, we consider some tools for the study of Pareto optimal points (POPs), and design an algorithm $\mathscr{A}_4$ for the construction of the set of POPs for a bi-criteria optimization problem.

1.4.2 Part II. Combinatorial Optimization Problems. Conventional Circuits

In the second part, we study nine combinatorial optimization problems. For each problem, we construct a conventional circuit without repetitions that represents all elements under consideration (the set of feasible solutions), describe at least two increasing or strictly increasing cost functions for this circuit, evaluate the number of operations and the time required by the algorithms $\mathscr{A}_1$, $\mathscr{A}_2$, and $\mathscr{A}_4$, consider an example of the problem instance, and discuss the results of experiments with randomly generated instances of the problem. This part contains nine chapters.

In Chap. 5, we study the problem of matrix chain multiplication.

In Chap. 6, we consider the global sequence alignment problem.

In Chap. 7, we discuss the problem of optimal paths in directed graphs.

In Chap. 8, we consider the binary search tree problem.

In Chap. 9, we study the problem of convex polygon triangulation.

In Chap. 10, we discuss the line breaking (text justification) problem.

In Chap. 11, we study the problem of one-dimensional clustering.

In Chap. 12, we consider the optimal bitonic tour problem.

In Chap. 13, we study the problem of segmented least squares.

1.4.3 Part III. Multi-Stage and Bi-Criteria Optimization of Matchings in Trees

In the third part, we propose an algorithm for counting matchings and algorithms for multi-stage and bi-criteria optimization of matchings relative to different weight functions that assign positive integers (weights) to edges of the tree. We did not find a conventional circuit without repetitions that represents the set of all matchings for an arbitrary tree. Instead of a circuit, we construct a labeled forest that describes the set of matchings, and design algorithms of matching optimization and counting based on the forest. We restrict ourselves to considering only bounds on the number of operations and do not consider the time complexity of the algorithms. The third part contains four chapters.

In Chap. 14, we discuss the problem of optimization of matchings in a tree and consider a labeled forest that describes the set of matchings.

In Chap. 15, we propose an algorithm for counting matchings and an algorithm for the optimization of matchings relative to a weight function.

In Chap. 16, we describe an additional tool for the study of Pareto optimal points (fusion of sets of POPs), propose an algorithm for bi-criteria optimization of matchings relative to two weight functions, and discuss the notion of totally optimal matching.

In Chap. 17, we consider an example of a tree and its matching optimization and discuss results of computer experiments with randomly generated trees and weight functions.

1.4.4 Part IV. Combinatorial Optimization Problems. Syntactical Circuits

In the fourth part, we study two combinatorial optimization problems for which we could not find conventional circuits without repetitions: optimization of matchings in trees and 0/1 knapsack problem. For each problem, we construct a conventional circuit with repetitions, transform this circuit into a syntactical circuit without repetitions, and define at least two increasing or strictly increasing cost functions for the syntactical circuit. We evaluate the number of operations and the time required by the algorithms $\mathscr{A}_1$, $\mathscr{A}_2$, and $\mathscr{A}_4$, consider an example of the problem instance, and discuss the results of experiments with randomly generated instances of the problem.

In Chap. 18, we study the problem of optimization of matchings in trees.

In Chap. 19, we consider the 0/1 knapsack problem.

1.5 Software Development

To demonstrate and validate our combinatorial optimization framework, the dedicated software has been developed. Most of the code has been implemented in Python, with supporting libraries partially written in C++. The software includes all algorithms introduced in this book, in particular, multi-stage optimization, counting elements described by the circuit, and bi-criteria optimization. The optimization software is implemented using the object-oriented approach. The main classes define the structure and elements of the circuit as well as they define optimization and supporting algorithms. There are problem-specific classes which inherit structure from one general class that defines basic functionality and structure of problems. For each combinatorial optimization problem, the recursion to build the circuit, cost functions, and input elements need to be defined independently. Additionally, the software includes classes used for experiments, in particular, plotting Pareto optimal points for bi-criteria optimization and parallel execution of multiple trails of optimization procedures.

1.6 Use of Book

The results presented in this book can be used for the study of multi-objective combinatorial optimization problems. If we are able to simulate a given problem by a conventional circuit without repetitions and by increasing or strictly increasing cost functions for this circuit, then we can apply the algorithms $\mathscr{A}_1$, $\mathscr{A}_2$, and $\mathscr{A}_4$ to investigate the problem. Otherwise, we can use syntactical circuits or try to create the multi-stage and bi-criteria optimization techniques for the considered problem from scratch as it is done for the optimization of matchings in a tree in Part III. This book can also be used in courses on combinatorial optimization for graduate students.

References

1. Abbasi, M., Paquete, L., Liefooghe, A., Pinheiro, M., Matias, P.: Improvements on bicriteria pairwise sequence alignment: algorithms and applications. Bioinformatics **29**(8), 996–1003 (2013)
2. AbouEisha, H., Amin, T., Chikalov, I., Hussain, S., Moshkov, M.: Extensions of Dynamic Programming for Combinatorial Optimization and Data Mining, Intelligent Systems Reference Library, vol. 146. Springer (2019)
3. AbuBekr, J., Chikalov, I., Hussain, S., Moshkov, M.: Sequential optimization of paths in directed graphs relative to different cost functions. In: Sato, M., Matsuoka, S., Sloot, P.M.A., van Albada, G.D., Dongarra, J. (eds.) International Conference on Computational Science, ICCS 2011, Nanyang Technological University, Singapore, June 1–3, 2011. Procedia Comput. Sci., vol. 4, pp. 1272–1277. Elsevier (2011)
4. Alnafie, M., Chikalov, I., Hussain, S., Moshkov, M.: Sequential optimization of binary search trees for multiple cost functions. In: Potanin, A., Viglas, T. (eds.) Seventeenth Computing: The Australasian Theory Symposium, CATS 2011, Perth, Australia, January 17–20, 2011. CRPIT, vol. 119, pp. 41–44. Australian Computer Society (2011)
5. Azad, M., Moshkov, M.: Multi-stage optimization of decision and inhibitory trees for decision tables with many-valued decisions. Eur. J. Oper. Res. **263**(3), 910–921 (2017)
6. Bazgan, C., Hugot, H., Vanderpooten, D.: Solving efficiently the 0–1 multi-objective knapsack problem. Comput. Oper. Res. **36**(1), 260–279 (2009)
7. Bellman, R.: The theory of dynamic programming. Bull. Amer. Math. Soc. **60**, 503–515 (1954)
8. Calvete, H.I., Mateo, P.M.: An approach for the network flow problem with multiple objectives. Comput. Oper. Res. **22**(9), 971–983 (1995)
9. Calvete, H.I., Mateo, P.M.: A sequential network-based approach for the multiobjective network flow problem with preemptive priorities. In: Tamiz, M. (ed.) Multi-Objective Programming and Goal Programming: Theories and Applications, pp. 74–86. Springer (1996)
10. Captivo, M.E., Clımaco, J., Figueira, J., Martins, E., Santos, J.L.: Solving bicriteria 0–1 knapsack problems using a labeling algorithm. Comput. Oper. Res. **30**(12), 1865–1886 (2003)
11. Chikalov, I., Hussain, S., Moshkov, M.: Relationships between depth and number of misclassifications for decision trees. In: Kuznetsov, S.O., Slezak, D., Hepting, D.H., Mirkin, B.G. (eds.) 13th International Conference Rough Sets, Fuzzy Sets, Data Mining and Granular Computing, RSFDGrC 2011, Moscow, Russia, June 25–27, 2011. Lecture Notes in Computer Science, vol. 6743, pp. 286–292. Springer (2011)

12. Chikalov, I., Hussain, S., Moshkov, M.: Sequential optimization of matrix chain multiplication relative to different cost functions. In: Cerná, I., Gyimóthy, T., Hromkovic, J., Jeffery, K.G., Královic, R.,Vukolic, M., Wolf, S. (eds.) 37th Conference on Current Trends in Theory and Practice of Computer Science, SOFSEM 2011, Nový Smokovec, Slovakia, January 22–28, 2011. Lecture Notes in Computer Science, vol. 6543, pp. 157–165. Springer (2011)
13. Chikalov, I., Hussain, S., Moshkov, M.: Relationships between number of nodes and number of misclassifications for decision trees. In: Yao, J., Yang, Y., Slowinski, R., Greco, S., Li, H., Mitra, S., Polkowski, L. (eds.) 8th International Conference Rough Sets and Current Trends in Computing, RSCTC 2012, Chengdu, China, August 17–20, 2012. Lecture Notes in Computer Science, vol. 7413, pp. 212–218. Springer (2012)
14. Chikalov, I., Hussain, S., Moshkov, M.: Bi-criteria optimization of decision trees with applications to data analysis. Eur. J. Oper. Res. **266**(2), 689–701 (2018)
15. Chikalov, I., Hussain, S., Moshkov, M., Odat, E.: Sequential optimization of global sequence alignments relative to different cost functions. In: ACM International Conference on Convergence and Hybrid Information Technology, ICHIT 2010, Daejeon, Korea, August 26–28, 2010. ACM (2010)
16. Cormen, T.H., Stein, C., Rivest, R.L., Leiserson, C.E.: Introduction to Algorithms, 2nd edn. The MIT Press and McGraw-Hill Book Company (2001)
17. Delort, C., Spanjaard, O.: Using bound sets in multiobjective optimization: Application to the biobjective binary knapsack problem. In: Festa, P. (ed.) Experimental Algorithms, pp. 253–265. Springer (2010)
18. Ehrgott, M.: Multicriteria Optimization. Springer (2000)
19. Figueira, J.R., Paquete, L., Simões, M., Vanderpooten, D.: Algorithmic improvements on dynamic programming for the bi-objective {0, 1} knapsack problem. Comp. Opt. and Appl. **56**(1), 97–111 (2013)
20. Floyd, R.W.: Algorithm 97: Shortest path. Commun. ACM **5**(6), 345 (1962)
21. Gatter, T., Giegerich, R., Saule, C.: Integrating Pareto optimization into dynamic programming. Algorithms **9**(1), 12 (2016). https://doi.org/10.3390/a9010012
22. Giegerich, R.: A declarative approach to the development of dynamic programming algorithms, applied to RNA folding. Report 98–02, Faculty of Technology, Bielefeld University (1998)
23. Giegerich, R., Kurtz, S., Weiller, G.F.: An algebraic dynamic programming approach to the analysis of recombinant DNA sequences. In: Workshop on Algorithmic Ascpects of Advanced Programming Languages, WAAAPL'99, Paris, France, September 30, pp. 77–88 (1999)
24. Giegerich, R., Steffen, P.: Pair evaluation algebras in dynamic programming. In: 21st Workshop of the GI-Fachgruppe Programming Languages and Computing Concepts, Bad Honnef, Germany, May 3–5, 2004, pp. 115–124 (2005)
25. Gilbert, P.D.: New results on planar triangulations. Master's thesis, University of Illinois at Urbana-Champaign (1979)
26. Godbole, S.: On efficient computation of matrix chain products. IEEE Trans. Comput. **22**(9), 864–866 (1973)
27. Grötschel, M., Lovász, L., Schrijver, A.: Geometric Algorithms and Combinatorial Optimization, Algorithms and Combinatorics, vol. 2. Springer (1988)
28. Hamacher, H.W., Labb, M., Nickel, S.: Multicriteria network location problems with sum objectives. Networks **33**(2), 79–92 (1999)
29. Hansen, P.: Bicriterion path problems. In: Fandel, G, Gal, T. (eds.) Multiple Criteria Decision Making Theory and Application. Lect. Notes Econ. Math., vol. 177, pp. 109–127. Springer (1980)
30. Holkner, A.: Global multiple objective line breaking. Master's thesis, RMIT University (2006)
31. Hussain, S.: Relationships among various parameters for decision tree optimization. In: Faucher, C., Jain, L.C. (eds.) Innovations in Intelligent Machines–4 Recent Advances in Knowledge Engineering. Studies in Computational Intelligence, vol. 514, pp. 393–410. Springer (2014)
32. Klamroth, K., Wiecek, M.M.: Dynamic programming approaches to the multiple criteria knapsack problem. Naval Res. Logistics **47**(1), 57–76 (2000)

33. Kleinberg, J., Tardos, E.: Algorithm Design. Addison-Wesley Longman Publishing Co., Inc. (2005)
34. Klincsek, G.T.: Minimal triangulations of polygonal domains. Ann. Discrete Math. **9**, 121–123 (1980)
35. Knuth, D.E.: Optimum binary search trees. Acta Inform. **1**(1), 14–25 (1971)
36. Knuth, D.E., Plass, M.F.: Breaking paragraphs into lines. Softw. Pract. Exp. **11**(11), 1119–1184 (1981)
37. Lupanov, O.B.: A method of circuit synthesis. Izvestia VUZ. Radiofizika **1**, 120–140 (1958)
38. Mankowski, M., Moshkov, M.: Dynamic programming bi-criteria combinatorial optimization. Discret. Appl. Math. **284**, 513–533 (2020)
39. Mankowski, M., Moshkov, M.: Extensions of dynamic programming for multi-stage combinatorial optimization. Theor. Comput. Sci. **844**, 106–132 (2020)
40. Mavrotas, G., Figueira, J.R., Antoniadis, A.: Using the idea of expanded core for the exact solution of bi-objective multi-dimensional knapsack problems. J. Global Optimization **49**(4), 589–606 (2011)
41. McKenzie, P., Wagner, K.W.: The complexity of membership problems for circuits over sets of natural numbers. Comput. Complexity **16**(3), 211–244 (2007)
42. Moshkov, M., Chikalov, I.: Sequential optimization of decision trees relatively different complexity measures. In: 6th International Conference Soft Computing and Distributed Processing, Rzeszòw, Poland, June 24–25, 2002, pp. 53–56 (2002)
43. Moshkov, M., Chikalov, I.: Consecutive optimization of decision trees concerning various complexity measures. Fundam. Inform. **61**(2), 87–96 (2004)
44. Needleman, S.B., Wunsch, C.D.: A general method applicable to the search for similarities in the amino acid sequence of two proteins. J. Mol. Biol. **48**(3), 443–453 (1970)
45. Ogryczak, W.: On the lexicographic minimax approach to location problems. Eur. J. Oper. Res. **100**(3), 566–585 (1997)
46. Raith, A., Ehrgott, M.: A comparison of solution strategies for biobjective shortest path problems. Comput. Oper. Res. **36**(4), 1299–1331 (2009)
47. Rong, A., Figueira, J.R.: Dynamic programming algorithms for the bi-objective integer knapsack problem. Eur. J. Oper. Res. **236**(1), 85–99 (2014)
48. Roytberg, M., Semionenkov, M., Tabolina, O.: Pareto-optimal alignment of biological sequences. Biofizika **44**(4), 581–594 (1999)
49. Saule, C., Giegerich, R.: Observations on the feasibility of exact Pareto optimization. In: F. Jossinet, Y. Ponty, J. Waldispühl (eds.) 1st Workshop on Computational Methods for Structural RNAs, CMSR 2014, Strasbourg, France, September 7, 2014, pp. 43–56. McGill University (2014). https://doi.org/10.15455/CMSR.2014.0004
50. Saule, C., Giegerich, R.: Pareto optimization in algebraic dynamic programming. Algorithms for Molecular Biology **10**, 22 (2015). https://doi.org/10.1186/s13015-015-0051-7
51. Schönhage, A., Strassen, V.: Schnelle multiplikation großer zahlen. Computing **7**(3–4), 281–292 (1971)
52. Shannon, C.E.: A symbolic analysis of relay and switching circuits. Trans. AIEE **57**(1–3), 59–98 (1938)
53. Sinuany-Stern, Z., Weiner, I.: The one dimensional cutting stock problem using two objectives. J. Oper. Res. Soc. **45**(2), 231–236 (1994)
54. Steffen, P., Giegerich, R.: Versatile and declarative dynamic programming using pair algebras. BMC Bioinform. **6**, 224 (2005)
55. Travers, S.D.: The complexity of membership problems for circuits over sets of integers. Theor. Comput. Sci. **369**(1–3), 211–229 (2006)
56. Visée, M., Teghem, J., Pirlot, M., Ulungu, E.L.: Two-phases method and branch and bound procedures to solve the bi-objective knapsack problem. J. Global Optimiz. **12**(2), 139–155 (1998)
57. Volgenant, A.: Solving some lexicographic multi-objective combinatorial problems. Eur. J. Oper. Res. **139**(3), 578–584 (2002)

58. Vollmer, H.: Introduction to Circuit Complexity - A Uniform Approach. Texts in Theoretical Computer Science. An EATCS Series. Springer (1999)
59. Vujošević, M.B., Stanojević, M.: A bicriterion Steiner tree problem on graph. Yugoslav J. Opera. Res. **13**(1), 25–33 (2003)
60. Wang, H., Song, M.: Ckmeans.1d.dp: Optimal k-means clustering in one dimension by dynamic programming. R J. **3**(2), 29–33 (2011)
61. Warshall, S.: A theorem on boolean matrices. J. ACM **9**(1), 11–12 (1962)

Part I
Circuits Without Repetitions. Main Tools

In this part, we consider notions and tools that are common for all chapters in Parts II and IV.

In Chap. 2, we discuss four basic notions: the notion of a circuit, the notion of a circuit without repetitions, the notion of a cost function for a circuit, and the notion of a syntactical circuit corresponding to a given circuit. We also consider four topics related to the algorithms over circuits: schemes of circuits, adaptive cost functions, bounds on the number of operations, and bounds on the time complexity.

In Chap. 3, we propose an algorithm $\mathscr{A}_1$, which can be used for multi-stage optimization of elements relative to a sequence of cost functions, and an algorithm $\mathscr{A}_2$ for finding the cardinalities of the sets of elements described by the initial circuit and its subcircuits obtained during the process of multi-stage optimization.

In Chap. 4, we consider some tools for the study of Pareto optimal points (POPs), and design an algorithm $\mathscr{A}_4$ for the construction of the set of POPs for a bi-criteria optimization problem.

Chapter 2
Circuits and Cost Functions

In this chapter, we consider the notion of a circuit and the notion of a cost function for this circuit. The circuit builds a set of elements for the optimization (a set of feasible solutions) from one-element sets attached to input nodes using the operation of union of sets attached to unifying nodes and functional operations attached to functional nodes. The cost function corresponds a real number (a cost) to each element. Our goal is to minimize the cost.

We mainly study so-called circuits without repetitions in which sets of elements corresponding to input and functional nodes are pairwise disjoint, and functions attached to functional nodes are injective. We consider only so-called increasing and strictly increasing cost functions for such circuits. Among these functions, we highlight the nonnegative and nonpositive integer cost functions. We also define the notions of an optimal element and a strongly optimal element, and prove that, for strictly increasing cost functions, these notions are equivalent.

We could not find for every problem a circuit without repetitions, which works directly with elements (we call such circuits conventional). Besides conventional circuits, we also consider so-called syntactical circuits that work with formulas representing elements. The syntactical circuits are circuits without repetitions. All algorithms considered in the next two chapters are applicable to the syntactical circuits. These algorithms are described on elements, while for the syntactical circuits, we need to interchange elements with formulas, which represent elements.

The algorithms for optimization and counting of elements considered in the next two chapters can work not only with circuits but also with schemes of circuits, obtained from circuits by the removal of one-element sets attached to input nodes, functions attached to functional nodes, and operations of union attached to unifying nodes. They usually work with so-called adapted cost functions given by formulas over a set of simple elementary operations. We can derive bounds on the number of elementary operations required by algorithms from the description of circuits and cost functions. We can transform the obtained bounds on the number of operations

M. Mankowski and M. Moshkov, *Dynamic Programming Multi-Objective Combinatorial Optimization*, Studies in Systems, Decision and Control 331,
https://doi.org/10.1007/978-3-030-63920-4_2

into bounds on the time complexity in the frameworks of the software and integer models of computation.

Note that the labeled directed acyclic graphs used in the book [1] for the representation and optimization of decision trees and decision rules can be transformed into circuits similar to ones studied in this chapter.

There is some similarity between the circuits considered in this chapter and combinatorial circuits that compute Boolean functions [3, 8, 10]. However, the circuits studied in this chapter are closer to $\{\cup, \cap, \bar{}, +, \times\}$-circuits computing (representing) sets of natural numbers [6] or integers [9]. This chapter is partially based on [4, 5].

2.1 Circuits

Let U be a finite set. A circuit S over U is a labeled directed acyclic graph (DAG) with three kinds of nodes:

- Input node v is a node without entering edges. It is labeled with a set $\{a_v\}$, where a_v is an element from U.
- Functional node v is a node with $k = k(v) \geq 1$ entering edges $e_1, \ldots, e_k$, which are labeled with numbers $1, \ldots, k$, respectively. This node is labeled with a function with k variables $f_v(x_1, \ldots, x_k)$ from $D_v(S)$ to U, where $D_v(S) \subseteq U^k$. For $i = 1, \ldots, k$, the entering edge with the number i corresponds to the variable x_i. We define the domain $D_v(S)$ of the function f_v later.
- Unifying node v is labeled with the symbol $\cup$. This node has $m \geq 1$ entering edges, which leave m pairwise different input or functional nodes. The number of entering edges can be different for different unifying nodes.

Some node of the circuit S is labeled with the symbol $*$ and is considered as the output of S.

For any node v of the circuit S, we define the set of elements $S(v) \subseteq U$ and, at the same time, define the domain $D_v(S)$ of the function f_v for each functional node v of S:

- If v is an input node of S, then $S(v) = \{a_v\}$.
- If v is a functional node of S with $k = k(v)$ entering edges $e_1, \ldots, e_k$, which are labeled with numbers $1, \ldots, k$ and leave nodes $v_1, \ldots, v_k$, respectively, then $S(v) = \{f_v(\delta_1, \ldots, \delta_k) : \delta_1 \in S(v_1), \ldots, \delta_k \in S(v_k)\}$, where f_v is a function from $D_v(S) = S(v_1) \times \cdots \times S(v_k)$ to U.
- If v is a unifying node of S with m entering edges $e_1, \ldots, e_m$, which leave nodes $v_1, \ldots, v_m$, respectively, then $S(v) = S(v_1) \cup \cdots \cup S(v_m)$.

We denote $V(S)$ the set of nodes of S, $V_i(S)$—the set of input nodes of S, $V_f(S)$—the set of functional nodes of S, and $V_u(S)$—the set of unifying nodes of S. Set $U(S) = \bigcup_{v \in V(S)} S(v)$. The function f_v, $v \in V_f(S)$, is called injective if $f_v(\alpha) \neq f_v(\beta)$ for any $\alpha, \beta \in D_v(S)$ such that $\alpha \neq \beta$.

We say that S is a circuit without repetitions if it satisfies the following conditions:

- $S(v_1) \cap S(v_2) = \emptyset$ for any nodes $v_1, v_2 \in V_i(S) \cup V_f(S)$ such that $v_1 \neq v_2$.
- The function f_v is injective for any functional node v of S.

Let S be a circuit without repetitions, v be a functional node of S with $k = k(v)$ entering edges $e_1, \ldots, e_k$, which are labeled with numbers $1, \ldots, k$ and leave nodes $v_1, \ldots, v_k$, and δ be an element from the set $S(v)$. Since the function f_v is injective, there exists only one tuple of elements $(\delta_1, \ldots, \delta_k) \in D_v(S) = S(v_1) \times \cdots \times S(v_k)$ such that $\delta = f_v(\delta_1, \ldots, \delta_k)$. We call the expression $f_v(\delta_1, \ldots, \delta_k)$ the representation of the element δ. Let now v be a unifying node of S with m entering edges $e_1, \ldots, e_m$, which leave nodes $v_1, \ldots, v_m$, and δ be an element from the set $S(v)$. Then there is exactly one $i \in \{1, \ldots, m\}$ such that $\delta \in S(v_i)$.

The set $S(v_*)$, where v_* is the output of S, will be considered as the set of elements for the definition of an optimization problem (the set of feasible solutions). The notion of a cost function will be defined in the next section.

2.2 Cost Functions

We now define the notion of a cost function Ψ for a circuit S without repetitions that corresponds to each element $\delta \in U(S)$ a real number $\Psi(\delta)$, which is considered as the cost of δ. The function Ψ is given by the function ψ, which corresponds to each element a_v, where v is an input node of S, a real number $\psi(a_v)$ and by functions $\psi_v : \mathbb{R}^{k(v)} \to \mathbb{R}$, $v \in V_f(S)$. Here $\mathbb{R}$ is the set of real numbers.

The value of $\Psi(\delta)$ for an element $\delta \in U(S)$ is defined by induction:

- If $\delta = a_v$, where v is an input node of S, then $\Psi(\delta) = \psi(a_v)$.
- If $\delta = f_v(\delta_1, \ldots, \delta_{k(v)}) \in S(v)$, where v is a functional node of S and $f_v(\delta_1, \ldots, \delta_k)$ is the representation of δ, then $\Psi(\delta) = \psi_v(\Psi(\delta_1), \ldots, \Psi(\delta_{k(v)}))$.

The function Ψ is called integer if the function ψ corresponds to each element a_v, where v is an input node of S, an integer $\psi(a_v)$ and each function ψ_v, $v \in V_f(S)$, has integer value if its inputs are integers.

We now define a partial order $\leq$ on the set $\mathbb{R}^k$. Let $\alpha = (\alpha_1, \ldots, \alpha_k)$, $\beta = (\beta_1, \ldots, \beta_k) \in \mathbb{R}^k$. Then $\alpha \leq \beta$ if $\alpha_1 \leq \beta_1, \ldots, \alpha_k \leq \beta_k$. We write $\alpha < \beta$ if $\alpha \leq \beta$ and $\alpha \neq \beta$. Let φ be a function from $\mathbb{R}^k$ to $\mathbb{R}$. We say that φ is increasing if $\varphi(\alpha) \leq \varphi(\beta)$ for any $\alpha, \beta \in \mathbb{R}^k$ such that $\alpha \leq \beta$. We say that φ is strictly increasing if $\varphi(\alpha) < \varphi(\beta)$ for any $\alpha, \beta \in \mathbb{R}^k$ such that $\alpha < \beta$. It is clear that if φ is strictly increasing, then φ is increasing.

We say that a cost function Ψ is increasing if, for any functional node v of S, the function ψ_v is increasing. We say that a cost function Ψ is strictly increasing if, for any functional node v of S, the function ψ_v is strictly increasing. We say that an integer cost function Ψ is a nonnegative integer cost function if, for any element $\delta \in U(S)$, the value of $\Psi(\delta)$ is a nonnegative integer. We say that an integer cost function Ψ is a nonpositive integer cost function if, for any element $\delta \in U(S)$, the value of $\Psi(\delta)$ is a nonpositive integer.

For each node v of S, we denote $\Psi^S(v) = \min\{\Psi(\delta) : \delta \in S(v)\}$. Let $v \in V(S)$ and $\delta \in S(v)$. We say that δ is optimal in $S(v)$ relative to Ψ if $\Psi(\delta) = \Psi^S(v)$. Note that if v is an input node, then the only element a_v from $S(v)$ is optimal in $S(v)$ relative to Ψ.

We define the notion of an element from $S(v)$, which is strongly optimal in $S(v)$ relative to Ψ by induction:

- Let v be an input node. Then the only element a_v from $S(v)$ is strongly optimal in $S(v)$ relative to Ψ.
- Let v be a functional node of S with $k = k(v)$ entering edges $e_1, \dots, e_k$, which are labeled with numbers $1, \dots, k$ and leave nodes $v_1, \dots, v_k$, δ be an element from the set $S(v)$, and $f_v(\delta_1, \dots, \delta_k)$ be the representation of δ. Then the element δ is strongly optimal in $S(v)$ relative to Ψ if δ is optimal in $S(v)$ relative to Ψ and, for $i = 1, \dots, k$, the element δ_i is strongly optimal in $S(v_i)$ relative to Ψ.
- Let v be a unifying node of S with m entering edges $e_1, \dots, e_m$, which leave nodes $v_1, \dots, v_m$, δ be an element from the set $S(v)$, $i \in \{1, \dots, m\}$, and $\delta \in S(v_i)$. Then the element δ is strongly optimal in $S(v)$ relative to Ψ if δ is optimal in $S(v)$ relative to Ψ and the element δ is strongly optimal in $S(v_i)$ relative to Ψ.

For each node v of S, we denote $S_\Psi^{opt}(v)$ the set of elements from $S(v)$, which are optimal in $S(v)$ relative to Ψ. We denote $S_\Psi^{s-opt}(v)$ the set of elements from $S(v)$ that are strongly optimal in $S(v)$ relative to Ψ. It is clear that $S_\Psi^{s-opt}(v) \subseteq S_\Psi^{opt}(v)$ for any node v of S. Note that $S_\Psi^{s-opt}(v) = \{a_v\} = S_\Psi^{opt}(v)$ for any input node v of S.

Proposition 2.1 *Let S be a circuit without repetitions and Ψ be a strictly increasing cost function for S. Then $S_\Psi^{s-opt}(v) = S_\Psi^{opt}(v)$ for any node v of S.*

Proof We know that $S_\Psi^{s-opt}(v) \subseteq S_\Psi^{opt}(v)$ for any node v of S. We now show by induction on nodes of S that $S_\Psi^{opt}(v) \subseteq S_\Psi^{s-opt}(v)$ for any node v of S. It is clear that $S_\Psi^{opt}(v) \subseteq S_\Psi^{s-opt}(v)$ for any input node v of S.

Let v be a functional node of S with $k = k(v)$ entering edges $e_1, \dots, e_k$, which are labeled with numbers $1, \dots, k$ and leave nodes $v_1, \dots, v_k$ such that $S_\Psi^{opt}(v_i) \subseteq S_\Psi^{s-opt}(v_i)$ for $i = 1, \dots, k$. Let $\delta \in S_\Psi^{opt}(v)$ and $f_v(\delta_1, \dots, \delta_k)$ be the representation of δ. We now show that $\delta_i \in S_\Psi^{opt}(v_i)$ for $i = 1, \dots, k$. Let us assume the contrary: there exists $i \in \{1, \dots, k\}$ such that $\delta_i \notin S_\Psi^{opt}(v_i)$. Then there is an element $\delta_i' \in S_\Psi^{opt}(v_i)$ for which $\Psi(\delta_i') < \Psi(\delta_i)$. It is clear that the element $\delta' = f_v(\delta_1, \dots, \delta_{i-1}, \delta_i', \delta_{i+1}, \dots, \delta_k)$ belongs to $S(v)$. Since Ψ is strictly increasing,

$$\begin{aligned}\Psi(\delta') &= \psi_v(\Psi(\delta_1), \dots, \Psi(\delta_{i-1}), \Psi(\delta_i'), \Psi(\delta_{i+1}), \dots, \Psi(\delta_k)) \\ &< \psi_v(\Psi(\delta_1), \dots, \Psi(\delta_{i-1}), \Psi(\delta_i), \Psi(\delta_{i+1}), \dots, \Psi(\delta_k)) \\ &= \Psi(\delta) ,\end{aligned}$$

but this is impossible because $\delta \in S_{\Psi}^{opt}(v)$. Therefore, $\delta_i \in S_{\Psi}^{opt}(v_i)$ for $i = 1, \ldots, k$ and, by the inductive hypothesis, $\delta_i \in S_{\Psi}^{s-opt}(v_i)$ for $i = 1, \ldots, k$. Taking into account that $\delta \in S_{\Psi}^{opt}(v)$, we obtain $\delta \in S_{\Psi}^{s-opt}(v)$. Therefore, $S_{\Psi}^{opt}(v) \subseteq S_{\Psi}^{s-opt}(v)$.

Let v be a unifying node of S with m entering edges $e_1, \ldots, e_m$, which leave nodes $v_1, \ldots, v_m$ such that $S_{\Psi}^{opt}(v_i) \subseteq S_{\Psi}^{s-opt}(v_i)$ for $i = 1, \ldots, m$. Let δ be an element from $S_{\Psi}^{opt}(v)$, $i \in \{1, \ldots, m\}$, and $\delta \in S(v_i)$. Since $\delta \in S_{\Psi}^{opt}(v)$, $\delta \in S_{\Psi}^{opt}(v_i)$. By the inductive hypothesis, $\delta \in S_{\Psi}^{s-opt}(v_i)$. Taking into account that $\delta \in S_{\Psi}^{opt}(v)$, we obtain $\delta \in S_{\Psi}^{s-opt}(v)$. Therefore, $S_{\Psi}^{opt}(v) \subseteq S_{\Psi}^{s-opt}(v)$.

As a result, we have $S_{\Psi}^{opt}(v) = S_{\Psi}^{s-opt}(v)$ for any node v of S. □

An edge-preserving subcircuit of a circuit S over U is a circuit T obtained from S by the removal of some edges entering unifying nodes such that each unifying node keeps at least one entering edge. By definition, S is an edge-preserving subcircuit of S. It is clear that $T(v) \subseteq S(v)$ for any node v of S, $U(T) \subseteq U(S)$ and, for any cost function for S, its restriction to $U(T)$ is a cost function for T. It is clear also that each edge-preserving subcircuit of a circuit without repetitions is a circuit without repetitions.

2.3 Syntactical Circuits

For some problems, we cannot find conventional (that work directly with elements) circuits without repetitions. Let S be an arbitrary circuit (with repetitions or without repetitions). We correspond to S so-called syntactical circuit $\hat{S}$, which is a circuit without repetitions. The circuit S describes a set of elements $S(v_*)$. The circuit $\hat{S}$ describes a set of formulas $\hat{S}(v_*)$ representing the same set of elements $S(v_*)$.

We correspond to each element a_v, $v \in V_i(S)$, an element symbol $\hat{a}_v$ such that $\hat{a}_{v_1} \neq \hat{a}_{v_2}$ for different input nodes v_1 and v_2. We correspond to each function f_v, $v \in V_f(S)$, a functional symbol $\hat{f}_v$ such that $\hat{f}_{v_1} \neq \hat{f}_{v_2}$ for different functional nodes v_1 and v_2. We also assume that the sets $\{\hat{a}_v : v \in V_i(S)\}$ and $\{\hat{f}_v : v \in V_f(S)\}$ are disjoint.

The circuit $\hat{S}$ is the same DAG as S but with different labels:

- If v is an input node of S labeled with the set $\{a_v\}$, then the corresponding node v of $\hat{S}$ is an input node of $\hat{S}$ labeled with the set $\{\hat{a}_v\}$.
- If v is a functional node of S labeled with the function $f_v(x_1, \ldots, x_k)$, then the corresponding node v of $\hat{S}$ is a functional node of $\hat{S}$ labeled with a function $\hat{f}_v(x_1, \ldots, x_k)$ described later.
- If v is a unifying node of S, then the corresponding node v of $\hat{S}$ is a unifying node of $\hat{S}$.
- If v is the output of S labeled with the symbol $*$, then the corresponding node v of $\hat{S}$ is the output of $\hat{S}$ labeled with the symbol $*$.

For any node v of the circuit $\hat{S}$, we define the set of formulas $\hat{S}(v)$ and, in the same time, define the domain $D_v(\hat{S})$ of the function $\hat{f}_v$ and the function $\hat{f}_v$ itself for each functional node v of $\hat{S}$:

- If v is an input node of $\hat{S}$, then $\hat{S}(v) = \{\hat{a}_v\}$.
- Let v be a functional node of $\hat{S}$ with $k = k(v)$ entering edges $e_1, \ldots, e_k$, which are labeled with numbers $1, \ldots, k$ and leave nodes $v_1, \ldots, v_k$, respectively. Then $\hat{f}_v$ is a function defined on the set $D_v(\hat{S}) = \hat{S}(v_1) \times \cdots \times \hat{S}(v_k)$. For any formulas $\delta_1 \in \hat{S}(v_1), \ldots, \delta_k \in \hat{S}(v_k)$, the value of the function $\hat{f}_v$ is the formula $\hat{f}_v(\delta_1, \ldots, \delta_k)$. Set $\hat{S}(v) = \{\hat{f}_v(\delta_1, \ldots, \delta_k) : \delta_1 \in \hat{S}(v_1), \ldots, \delta_k \in \hat{S}(v_k)\}$.
- If v is a unifying node of $\hat{S}$ with m entering edges $e_1, \ldots, e_m$, which leave nodes $v_1, \ldots, v_m$, respectively, then $\hat{S}(v) = \hat{S}(v_1) \cup \cdots \cup \hat{S}(v_m)$.

The circuit $\hat{S}$ is a circuit over the set of formulas $\bigcup_{v \in V(\hat{S})} \hat{S}(v)$. We know that symbols $\hat{a}_{v_1}, \hat{a}_{v_2}, \hat{f}_{v_3}, \hat{f}_{v_4}$ are pairwise different for different nodes $v_1, v_2 \in V_i(\hat{S})$ and $v_3, v_4 \in V_f(\hat{S})$. Therefore, $\hat{S}(v_1) \cap \hat{S}(v_2) = \emptyset$ for any nodes $v_1, v_2 \in V_i(\hat{S}) \cup V_f(\hat{S})$ such that $v_1 \neq v_2$. It is clear that, for any two tuples of formulas $(\delta_1, \ldots, \delta_k)$ and $(\sigma_1, \ldots, \sigma_k)$ such that $(\delta_1, \ldots, \delta_k) \neq (\sigma_1, \ldots, \sigma_k)$, the formulas $\hat{f}_v(\delta_1, \ldots, \delta_k)$ and $\hat{f}_v(\sigma_1, \ldots, \sigma_k)$ are different. Therefore, the function $\hat{f}_v$ is injective for any functional node v of $\hat{S}$. Thus, the circuit $\hat{S}$ is a circuit without repetitions.

Let Ψ be a cost function for the circuit $\hat{S}$ given by the function ψ, which corresponds to each symbol of element $\hat{a}_v$, where v is an input node of $\hat{S}$, a real number $\psi(\hat{a}_v)$, and by functions $\psi_v : \mathbb{R}^{k(v)} \to \mathbb{R}$, $v \in V_f(\hat{S})$.

The value of $\Psi(\delta)$ for a formula $\delta \in \bigcup_{v \in V(\hat{S})} \hat{S}(v)$ is defined by induction in the usual way:

- If $\delta = \hat{a}_v$, where v is an input node of $\hat{S}$, then $\Psi(\delta) = \psi(\hat{a}_v)$.
- If $\delta = \hat{f}_v(\delta_1, \ldots, \delta_{k(v)}) \in \hat{S}(v)$, where v is a functional node of $\hat{S}$, then $\Psi(\delta) = \psi_v(\Psi(\delta_1), \ldots, \Psi(\delta_{k(v)}))$.

Let δ be a formula from $\bigcup_{v \in V(\hat{S})} \hat{S}(v)$. If in this formula we set a_v instead of $\hat{a}_v$ for each $v \in V_i(\hat{S})$ and f_v instead of $\hat{f}_v$ for each $v \in V_f(\hat{S})$, we obtain a formula, which represents an element from the set $\bigcup_{v \in V(S)} S(v)$. We denote this element $a(\delta)$. One can show that $\{a(\delta) : \delta \in \hat{S}(v)\} = S(v)$ for any node v of the circuit $\hat{S}$.

In this book, for syntactical circuits, we consider only cost functions such that $\Psi(\delta_1) = \Psi(\delta_2)$ for any formulas δ_1 and δ_2 from $\hat{S}(v_*)$ for which $a(\delta_1) = a(\delta_2)$. We call such cost functions proper cost functions for the circuit $\hat{S}$ relative to the circuit S. For proper cost functions, instead of multi-stage or bi-criteria optimization on the set $S(v_*)$ we can consider multi-stage or bi-criteria optimization on the set $\hat{S}(v_*)$. To this end, we can use algorithms $\mathcal{A}_1$ (see Chap. 3) and $\mathcal{A}_4$ (see Chap. 4). In the description of these algorithms we say about elements. For the syntactical circuits, elements are formulas. The same situation is with the algorithm $\mathcal{A}_2$ (see Chap. 3), which allows us to count the number of formulas in the set $\hat{S}(v_*)$. Note that in the general case the cardinality of the set $\hat{S}(v_*)$ is greater than the cardinality of the set $S(v_*)$.

We do not use in this book the possibility to have a different cost for different formulas in $\hat{S}(v_*)$ representing the same element from $S(v_*)$. However, this possibility is vital in applications for which the cost of elements can depend on their representations.

2.4 Algorithms Over Circuits

2.4.1 Schemes of Circuits

In the next two chapters, we consider algorithms $\mathscr{A}_1$, $\mathscr{A}_2$, and $\mathscr{A}_4$ for optimization, counting, and bi-criteria optimization of elements, which work with circuits without repetitions. Let S be such a circuit. The elements a_v from the sets attached to the input nodes v of S and the functions f_v attached to the functional nodes v of S can be complicated. Fortunately, the algorithms $\mathscr{A}_1$, $\mathscr{A}_2$, and $\mathscr{A}_4$ do not use the functions f_v and do not use the elements a_v directly.

We remove the sets $\{a_v\}$ attached to the input nodes v of S, the functions f_v attached to the functional nodes v of S, and the symbols $\cup$ attached to the unifying nodes of S. We will not remove natural numbers attached to the edges of the scheme S. We label each node of S with its type (a letter from the set $\{i, f, u\}$, where i means an input node, f means a functional node, and u means a unifying node) and its unique name. As a result, we obtain a labeled directed acyclic graph, which will be called the scheme of the circuit S.

The algorithm $\mathscr{A}_2$ for counting elements described by the circuit S does not use elements a_v and functions f_v. It will work correctly when we use the scheme of S instead of the circuit S.

Let Ψ be a cost function for the circuit S given by the function ψ, which corresponds to each element a_v, where v is an input node of S, a real number $\psi(a_v)$ and by functions $\psi_v : \mathbb{R}^{k(v)} \rightarrow \mathbb{R}$, $v \in V_f(S)$.

The algorithm $\mathscr{A}_1$ for the optimization of elements described by the circuit S relative to the cost function Ψ does not use functions f_v. This algorithm does not use elements a_v directly, but only values $\psi(a_v)$. It will work correctly when we use the scheme of S instead of the circuit S and can describe the values $\psi(a_v)$, $v \in V_i(S)$.

Let Ψ, Φ be cost functions for S given, respectively, by the functions ψ, ψ_v and φ, φ_v, $v \in V_f(S)$.

The algorithm $\mathscr{A}_4$ for bi-criteria optimization of elements described by the circuit S relative to the cost functions Ψ, Φ does not use functions f_v. This algorithm does not use elements a_v directly, but only values $\psi(a_v)$ and $\varphi(a_v)$. It will work correctly when we use the scheme of S instead of the circuit S and can describe the values $\psi(a_v)$ and $\varphi(a_v)$, $v \in V_i(S)$.

The scheme of the circuit S can be given by the array of all nodes of S. Each node in this array is represented by pair (name, type) when the name is unique for every node. Let v be a node of S with m entering edges $e_1, \ldots, e_m$ leaving nodes

$v_1, \ldots, v_m$, respectively. The node v in the considered array has a list (name of node v_1, q_1), ..., (name of node v_m, q_m) associated with it. For $i = 1, \ldots, m$, q_i is equal to the natural number attached to the edge e_i. If the edge e_i is not labeled, then $q_i = \lambda$.

Usually, the time complexity of scheme construction is linear depending on the number of its nodes and edges. We will mention special cases when we need to take additional steps to construct the scheme. In particular, when it is necessary to check if the considered node or edge belongs to the scheme.

2.4.2 Adapted Cost Functions

Let Ψ be a cost function for the circuit S given by the function ψ and by functions ψ_v, $v \in V_f(S)$.

In all the cases considered in this book, the functions ψ_v for the functional nodes v of the circuit S are either equal to variables, i.e., $\psi_v(x) = x$ (in this case, we do not need any operations to compute the value of ψ_v), or represented by enough simple formulas over a set of elementary operations B using variables and some numeric parameters of the problem. Elementary operations from the set B are of the kind addition $x + y$, subtraction $x - y$, multiplication $x \cdot y$, division x/y, minimum $\min(x, y)$, maximum $\max(x, y)$, square root $\sqrt{x}$, ceiling $\lceil x \rceil$, floor $\lfloor x \rfloor$, absolute value $|x|$. The values $\psi(a_v)$ for the input nodes v of the circuit S are either given directly, for example, $\psi(a_v) = 0$ (in this case, we do not need any operations to compute the value of ψ), or represented by enough simple formulas over the set of elementary operations B using some numeric parameters of the problem. We will call such cost functions adapted cost functions.

An adapted cost function can be given by the array of input nodes v with constants or formulas representing the values $\psi(a_v)$, and by the array of functional nodes v with variables or formulas representing functions ψ_v.

The work of the algorithms $\mathcal{A}_1$, $\mathcal{A}_2$, and $\mathcal{A}_4$ for optimization, counting, and bi-criteria optimization of elements is described for the circuits without repetitions and for arbitrary cost functions for these circuits.

However, for each of the combinatorial optimization problems considered in the book, the algorithms $\mathcal{A}_1$, $\mathcal{A}_2$, and $\mathcal{A}_4$ really work with the schemes of circuits instead of the circuits and with adapted cost functions.

2.4.3 Bounds on Number of Operations

Each of the considered in the book problems has inherent dimensions—one or two natural numbers that describe the structure of problem instance. The problem instance includes polynomial in dimensions number of numeric parameters. Each problem dimension is at most the number of such parameters plus one. For the problem of matrix chain multiplication, the dimension is the number n of matrices and the numeric parameters are $n + 1$ natural numbers $m_0, m_1, \ldots, m_n$ that define sizes $m_0 \times m_1, m_1 \times m_2, \ldots, m_{n-1} \times m_n$ of these matrices. For the global sequence alignment problem, the dimensions are the lengths m and n of sequences. An instance of this problem has $m + n + mn$ numeric parameters that describe the costs of alignment for symbols and alignment for symbols and gaps.

Theorems 3.1, 3.2, and 4.1 give upper bounds on the number of operations made by the algorithms $\mathscr{A}_1$, $\mathscr{A}_2$, and $\mathscr{A}_4$, respectively. For $\mathscr{A}_1$, the operations are computations of values of ψ and ψ_v, $v \in V_f(S)$, and comparisons. For $\mathscr{A}_2$, the operations are additions and multiplications. For $\mathscr{A}_4$, the operations are computations of values of ψ, φ, φ_v, and ψ_v, $v \in V_f(S)$, and comparisons. The operation of comparison $x : y$ has three values $x < y$, $x = y$, and $x > y$. This operation allows us to find values $\min(x, y)$, $\max(x, y)$, to check if $x > y$ or $x \geq y$, etc.

For adapted cost functions, we deal with formulas over a set of elementary operations B and we can recalculate the upper bounds given by Theorems 3.1, 3.2, and 4.1 into upper bounds on the number of elementary operations from a set B^+, which contains all operations from B, the operation of comparison $x : y$, and some operations from the set $\{x + y, x \cdot y\}$.

As a result, we obtain two kinds of upper bounds on the number of operations from B^+: polynomial depending on the problem dimensions and polynomial depending on the problem dimensions and some numeric parameters of the problem (for example, the maximum weight of an edge in the problem of optimization of matching in trees, or $\max\{m_0, m_1, \ldots, m_n\}$ for the problem of matrix chain multiplication). In the first case, we say about the polynomial number of operations, in the second case—about the pseudo-polynomial number of operations.

2.4.4 Bounds on Time Complexity

To derive bounds on the time complexity of the considered algorithms from bounds on the number of operations, we need to fix a model of computation. We study two such models: the software model and the integer model [4, 5].

The software model is closely related to the conventional software when each of the considered operations requires a small constant number of clock cycles and we can use both variables for integers and variables for floating-point numbers. This model assumes that we are working with finite sets of numbers, for example, with single-precision (32-bit) floating-point numbers and 32-bit signed integers. In particular, for one of the Intel's processors of Broadwell architecture, single-precision operations $x + y$, $x - y$, $x \cdot y$, $\min(x, y)$, and $\max(x, y)$ require three clock cycles each, x/y requires 13–17, and $\sqrt{x}$ requires 19–35 clock cycles. For 32-bit signed integers, operations $x + y$ and $x - y$ require one clock cycle each, and the operation $x \cdot y$ requires ten clock cycles.

For the software model, from the existence of polynomial or pseudo-polynomial bounds on the number of operations, it follows that the considered algorithms have polynomial or pseudo-polynomial time complexities. For the algorithm $\mathscr{A}_4$, to keep upper bounds true, we need to use variables for integers when we deal with non-negative integer and nonpositive integer cost functions. If we have an upper bound A on the number of operations for an algorithm, then $O(A)$ is the upper bound on the time complexity of the algorithm in the framework of the software model of computation. All obtained bounds depend on problem dimensions and (in the case of pseudo-polynomial bounds) on some numeric parameters of the problem.

The integer model assumes that all inputs are integers of arbitrary size, and each operation transforms a tuple of integers into an integer. In this case, the time complexity of an operation depends on the maximum number of bits t in the binary representation of integers at the operation inputs. For example, minimum, maximum, comparison, addition, and subtraction require $O(t)$ and the multiplication requires $O(t \cdot \log t \cdot \log\log t)$ time [7].

For the integer model for each of the considered cases with an exception described later, when the instance of a problem is given by integer parameters in binary representation and we use integer cost functions, we will transform polynomial bounds on the number of operations depending on the problem dimensions into polynomial bounds on the time complexity of algorithms depending on the length of the problem representation. We will also transform pseudo-polynomial bounds on the number of operations (polynomial depending on the problem dimensions and some numeric parameters of the problem) into pseudo-polynomial bounds on the time complexity of algorithms (polynomial depending on the length of the problem representation and some numeric parameters of the problem). Note that for each of the considered problems, the problem dimensions are at most the length of the problem representation.

To obtain bounds on the time complexity, it will be necessary for us to evaluate the maximum absolute value of numbers used during the work of the algorithms $\mathcal{A}_1$, $\mathcal{A}_2$, $\mathcal{A}_4$, and the modification of the algorithm $\mathcal{A}_4$. The ways how to evaluate the maximum absolute value of the used numbers are described in Remarks 3.2, 3.4, 4.4, and 4.5 for the algorithms $\mathcal{A}_1$, $\mathcal{A}_2$, $\mathcal{A}_4$, and the modification of the algorithm $\mathcal{A}_4$, respectively. In some cases, we need an additional study related to the processing of input and functional nodes and the calculation of so-called prefix sum sequences. The only issue is with the evaluation of the time complexity of the algorithm $\mathcal{A}_2$ for syntactical circuits. In this case, we cannot find reasonable upper bounds on the number of formulas described by circuits. As a result, we cannot evaluate the size of numbers during the work of the algorithm $\mathcal{A}_2$.

If we have an upper bound A on the number of operations of comparison, minimum, maximum, multiplication, addition, and subtraction required by an algorithm, and m is an upper bound on the absolute values of numbers used during the work of the algorithm, then $O(A \cdot t \cdot \log t \cdot \log\log t)$, where $t = \log_2 m$, is the upper bound on the time complexity of the considered algorithm in the framework of the integer model of computation. If we do not use the operation of multiplication, then $O(A \cdot t)$ is the upper bound on the time complexity of the considered algorithm.

This way to obtain bounds on the time complexity of algorithms (evaluation of the number of operations and the size of used integers) is close to the notion of strongly polynomial time [2].

References

1. AbouEisha, H., Amin, T., Chikalov, I., Hussain, S., Moshkov, M.: Extensions of Dynamic Programming for Combinatorial Optimization and Data Mining, Intelligent Systems Reference Library, vol. 146. Springer (2019)
2. Grötschel, M., Lovász, L., Schrijver, A.: Geometric Algorithms and Combinatorial Optimization, Algorithms and Combinatorics, vol. 2. Springer (1988)
3. Lupanov, O.B.: A method of circuit synthesis. Izvestia VUZ. Radiofizika **1**, 120–140 (1958)
4. Mankowski, M., Moshkov, M.: Dynamic programming bi-criteria combinatorial optimization. Discret. Appl. Math. **284**, 513–533 (2020)
5. Mankowski, M., Moshkov, M.: Extensions of dynamic programming for multi-stage combinatorial optimization. Theor. Comput. Sci. **844**, 106–132 (2020)
6. McKenzie, P., Wagner, K.W.: The complexity of membership problems for circuits over sets of natural numbers. Comput. Complexity **16**(3), 211–244 (2007)
7. Schönhage, A., Strassen, V.: Schnelle multiplikation großer zahlen. Computing **7**(3–4), 281–292 (1971)
8. Shannon, C.E.: A symbolic analysis of relay and switching circuits. Trans. AIEE **57**(1–3), 59–98 (1938)
9. Travers, S.D.: The complexity of membership problems for circuits over sets of integers. Theor. Comput. Sci. **369**(1–3), 211–229 (2006)
10. Vollmer, H.: Introduction to Circuit Complexity - A Uniform Approach. Texts in Theoretical Computer Science. An EATCS Series. Springer (1999)

Chapter 3
Multi-Stage Optimization and Counting Optimal Elements

In this chapter, we propose an optimization procedure (an algorithm $\mathscr{A}_1$), which for a given circuit without repetitions and cost function for this circuit, transforms the circuit into its subcircuit. For strictly increasing cost functions, the set of elements described by the subcircuit coincides with the set of optimal elements relative to the considered cost function. For increasing cost functions, the set of elements described by the subcircuit is a nonempty subset of the set of optimal elements.

The algorithm $\mathscr{A}_1$ can be used for multi-stage optimization of elements relative to a sequence of cost functions. We also propose an algorithm $\mathscr{A}_2$ for finding the cardinalities of the sets of elements described by the initial circuit and its subcircuits obtained during the process of multi-stage optimization.

In a sense, similar algorithms were considered in [1, 3] for multi-stage optimization of decision and inhibitory trees, decision rules, decision rule systems, element partition trees, parenthesizations for matrix chain multiplication, global sequence alignments, paths in directed graphs, and binary search trees. Note that each problem was studied separately with its own problem-specific algorithms and proofs.

One of the applications of the created techniques is the study of the existence of elements that are optimal relative to several cost functions at the same time. We call such elements totally optimal relative to the considered cost functions and investigate their existence for various combinatorial optimization problems. Totally optimal decision trees, decision rules, decision rule systems, and element partition trees were studied earlier in [1, 2, 4]. This chapter is based on [5].

3.1 Algorithm for Optimization of Elements

Let S be a circuit without repetitions and Ψ be an increasing cost function for S. We now describe a procedure for optimization of elements described by S relative to the cost function Ψ—the algorithm $\mathscr{A}_1$. This algorithm assigns to each node v of S the

M. Mankowski and M. Moshkov, *Dynamic Programming Multi-Objective Combinatorial Optimization*, Studies in Systems, Decision and Control 331,
https://doi.org/10.1007/978-3-030-63920-4_3

number $c(v) = \Psi^S(v) = \min\{\Psi(\delta) : \delta \in S(v)\}$ and removes some edges entering unifying nodes of S. As a result, we obtain an edge-preserving subcircuit S^Ψ of S such that $S^\Psi(v) = S_\Psi^{s-opt}(v)$ for any node v of S.

Algorithm $\mathscr{A}_1$ (procedure of optimization of elements).
Input: A circuit S without repetitions and an increasing cost function Ψ for S given by functions ψ and ψ_v, $v \in V_f(S)$.
Output: An edge-preserving subcircuit S^Ψ of S with a number $c(v)$ assigned to each node v.

1. If all nodes of S are processed, then return the obtained circuit as S^Ψ.
2. Otherwise, choose a node v of S that is not processed yet and which is either an input node or a node for which all parents are processed.
3. If v is an input node, then assign to it the number $c(v) = \psi(a_v)$, label it as processed, and proceed to step 1.
4. If v is a functional node with $k = k(v)$ entering edges, which are labeled with numbers $1, \ldots, k$ and leave the nodes $v_1, \ldots, v_k$, respectively, then assign to v the number $c(v) = \psi_v(c(v_1), \ldots, c(v_k))$, label v as processed, and proceed to step 1.
5. If v is a unifying node with m entering edges $e_1, \ldots, e_m$ leaving the nodes $v_1, \ldots, v_m$, respectively, then assign to v the number $c(v) = \min\{c(v_1), \ldots, c(v_m)\}$ and remove all edges e_i, $i \in \{1, \ldots, m\}$, for which $c(v_i) > c(v)$. Label v as processed and proceed to step 1.

Remark 3.1 Note that the algorithm $\mathscr{A}_1$ can work with the schemes of circuits instead of the circuits and with adapted cost functions (see Sect. 2.4). In this case, the output of the algorithm $\mathscr{A}_1$ is the scheme of an edge-preserving subcircuit of the initial circuit.

Remark 3.2 Let S be a circuit without repetitions and Ψ be an increasing integer cost function for S given by functions ψ and ψ_v, $v \in V_f(S)$. Denote $N_\Psi(S) = \max\{|\Psi(\delta)| : \delta \in U(S)\}$, where $U(S) = \bigcup_{v \in V(S)} S(v)$. It is clear that, for each node v of S, the absolute value of the number $c(v) = \Psi^S(v) = \min\{\Psi(\delta) : \delta \in S(v)\}$ is at most $N_\Psi(S)$. Therefore, for the algorithm $\mathscr{A}_1$, the number $N_\Psi(S)$ is an upper bound on the absolute values of the numbers at the inputs of operations ψ_v related to the functional nodes and at the inputs of comparisons related to the unifying nodes. When a function ψ_v is given by a formula containing more than one elementary operation, we need to evaluate the absolute values of the numbers at the inputs of each elementary operation. For all cases studied in this book, it is easy to show that these absolute values are also at most $N_\Psi(S)$ or are at most some upper bound on $N_\Psi(S)$ considered in the corresponding chapter. In particular, let the values of the function ψ be nonnegative and functions ψ_v be given by formulas containing only operations $x + y$, $\max(x, y)$ and nonnegative numeric parameters. Let us assume that the absolute value of at least one number at an input of an elementary operation from the formula for ψ_v is greater than $N_\Psi(S)$. Then the absolute value of this formula is greater than $N_\Psi(S)$, which is impossible. The same situation is in the case when values of the function ψ are nonpositive and functions ψ_v are given by formulas containing only operations $x + y$, $\min(x, y)$ and nonpositive numeric parameters. The elementary operations related to the input nodes should be considered separately and only in the case when values of ψ are represented not by constants, but by formulas.

Theorem 3.1 *Let S be a circuit without repetitions and Ψ be an increasing cost function for S given by functions ψ and ψ_v, $v \in V_f(S)$. Then, for any node v of S, $c(v) = \Psi^S(v)$ and $S^\Psi(v) = S_\Psi^{s-opt}(v)$. For any node v of S, the algorithm $\mathcal{A}_1$ computes one time a value of ψ if v is an input node, computes one time a value of ψ_v if v is a functional node, and makes at most $2m - 1$ comparisons if v is a unifying node with m entering edges.*

Proof First, we prove by induction on nodes of S that $c(v) = \Psi^S(v)$ for any node v of S. Let v be an input node. Then $S(v) = \{a_v\}$ and, evidently, $c(v) = \psi(a_v) = \Psi^S(v)$. Let v be a functional node with $k = k(v)$ entering edges $e_1, \ldots, e_k$, which are labeled with numbers $1, \ldots, k$ and leave nodes $v_1, \ldots, v_k$ such that $c(v_1) = \Psi^S(v_1), \ldots, c(v_k) = \Psi^S(v_k)$. Since Ψ is an increasing cost function,

$$\Psi^S(v) = \psi_v(\Psi^S(v_1), \ldots, \Psi^S(v_k)) .$$

Using the inductive hypothesis, we obtain that $c(v) = \Psi^S(v)$. Let v be a unifying node with m entering edges $e_1, \ldots, e_m$ leaving nodes $v_1, \ldots, v_m$ such that $c(v_1) = \Psi^S(v_1), \ldots, c(v_m) = \Psi^S(v_m)$. It is clear that $\Psi^S(v) = \min\{\Psi^S(v_1), \ldots, \Psi^S(v_m)\}$. Using the inductive hypothesis, we obtain that $c(v) = \Psi^S(v)$.

We now prove by induction on nodes of S that $S^\Psi(v) = S_\Psi^{s-opt}(v)$ for any node v of S. Let v be an input node of S. Then $S^\Psi(v) = \{a_v\} = S_\Psi^{s-opt}(v)$.

Let v be a functional node with $k = k(v)$ entering edges $e_1, \ldots, e_k$, which are labeled with numbers $1, \ldots, k$ and leave nodes $v_1, \ldots, v_k$ such that

$$S^\Psi(v_1) = S_\Psi^{s-opt}(v_1), \ldots, S^\Psi(v_k) = S_\Psi^{s-opt}(v_k) .$$

By definition, $S^\Psi(v) = \{f_v(\delta_1, \ldots, \delta_k) : \delta_1 \in S^\Psi(v_1), \ldots, \delta_k \in S^\Psi(v_k)\}$. Let us consider an arbitrary element δ from $S^\Psi(v)$ with the representation $f_v(\delta_1, \ldots, \delta_k)$. Using the inductive hypothesis and the fact that Ψ is increasing, we obtain that δ is optimal in $S(v)$ relative to Ψ and, hence, is strongly optimal in $S(v)$ relative to Ψ. Therefore, $S^\Psi(v) \subseteq S_\Psi^{s-opt}(v)$. Let $\delta \in S_\Psi^{s-opt}(v)$. Then $\delta = f_v(\delta_1, \ldots, \delta_k)$, where $\delta_i \in S(v_i)$ for $i = 1, \ldots, k$. Since $\delta \in S_\Psi^{s-opt}(v)$, $\delta_i \in S_\Psi^{s-opt}(v_i)$ and, by the inductive hypothesis, $\delta_i \in S^\Psi(v_i)$ for $i = 1, \ldots, k$. Therefore, $\delta \in S^\Psi(v)$, $S_\Psi^{s-opt}(v) \subseteq S^\Psi(v)$, and $S_\Psi^{s-opt}(v) = S^\Psi(v)$.

Let v be a unifying node with m entering edges $e_1, \ldots, e_m$ leaving nodes $v_1, \ldots, v_m$ such that $S^\Psi(v_1) = S_\Psi^{s-opt}(v_1), \ldots, S^\Psi(v_m) = S_\Psi^{s-opt}(v_m)$. It is clear that $S^\Psi(v) = \bigcup_{\substack{i=1,\ldots,m \\ \Psi^S(v_i)=\Psi^S(v)}} S^\Psi(v_i)$. Let $\delta \in S^\Psi(v)$. According to the inductive hypothesis, δ is optimal in $S(v)$ relative to Ψ and δ is strongly optimal in $S(v_i)$ relative to Ψ for some $i \in \{1, \ldots, m\}$ such that $\Psi^S(v_i) = \Psi^S(v)$. Therefore, $\delta \in S_\Psi^{s-opt}(v)$. Let

$\delta \in S_{\Psi}^{s-opt}(v)$. Then $\Psi(\delta) = \Psi^S(v)$ and $\delta \in S_{\Psi}^{s-opt}(v_i)$ for some $i \in \{1, \ldots, m\}$. We have $\Psi^S(v_i) = \Psi(\delta) = \Psi^S(v)$. According to the inductive hypothesis, $\delta \in S^{\Psi}(v_i)$. Therefore, $\delta \in S^{\Psi}(v)$ and $S^{\Psi}(v) = S_{\Psi}^{s-opt}(v)$.

A simple analysis of the algorithm $\mathscr{A}_1$ shows that, for any node v of S, the algorithm $\mathscr{A}_1$ computes one time a value of ψ if v is an input node, computes one time a value of ψ_v if v is a functional node, and makes at most $2m - 1$ comparisons if v is a unifying node with m entering edges. □

The next statement follows immediately from Proposition 2.1 and Theorem 3.1.

Corollary 3.1 *Let S be a circuit without repetitions and Ψ be a strictly increasing cost function for S. Then $S^{\Psi}(v) = S_{\Psi}^{opt}(v)$ for any node v of S.*

3.2 Multi-Stage Optimization and Totally Optimal Elements

Let S be a circuit without repetitions and v_* be the output of S. The algorithm $\mathscr{A}_1$ can be used for multi-stage optimization of elements from $S(v_*)$ relative to a sequence of strictly increasing cost functions $\Psi_1, \Psi_2, \ldots$.

First, we apply the algorithm $\mathscr{A}_1$ to the circuit S and the cost function Ψ_1. As a result, we obtain an edge-preserving subcircuit S^{Ψ_1} of the circuit S. From Corollary 3.1 it follows that $S^{\Psi_1}(v_*) = S_{\Psi_1}^{opt}(v_*)$, i.e., the set $S^{\Psi_1}(v_*)$ coincides with the set of all elements from $S(v_*)$ that are optimal in $S(v_*)$ relative to Ψ_1. Next, we apply the algorithm $\mathscr{A}_1$ to the circuit S^{Ψ_1} and the cost function Ψ_2. As a result, we obtain an edge-preserving subcircuit S^{Ψ_1,Ψ_2} of the circuit S^{Ψ_1}. From Corollary 3.1 it follows that $S^{\Psi_1,\Psi_2}(v_*) = (S^{\Psi_1})_{\Psi_2}^{opt}(v_*)$, i.e., the set $S^{\Psi_1,\Psi_2}(v_*)$ coincides with the set of all elements from $S^{\Psi_1}(v_*)$ that are optimal in $S^{\Psi_1}(v_*)$ relative to Ψ_2, etc.

If one of the cost functions Ψ_i, $i > 1$, is increasing and not strictly increasing, then, by Theorem 3.1, the set $S^{\Psi_1,\ldots,\Psi_i}(v_*)$ coincides with the set $(S^{\Psi_1,\ldots,\Psi_{i-1}})_{\Psi_i}^{s-opt}(v_*)$ of all elements from $S^{\Psi_1,\ldots,\Psi_{i-1}}(v_*)$ that are strongly optimal in $S^{\Psi_1,\ldots,\Psi_{i-1}}(v_*)$ relative to Ψ_i. In the general case, this set is a subset of the set $(S^{\Psi_1,\ldots,\Psi_{i-1}})_{\Psi_i}^{opt}(v_*)$ of all elements from $S^{\Psi_1,\ldots,\Psi_{i-1}}(v_*)$ that are optimal in $S^{\Psi_1,\ldots,\Psi_{i-1}}(v_*)$ relative to Ψ_i.

Let $\Psi_1, \ldots, \Psi_m$ be cost functions for S and δ be an element from the set $S(v_*)$. We say that δ is totally optimal in $S(v_*)$ relative to $\Psi_1, \ldots, \Psi_m$ if $\Psi_i(\delta) = \Psi_i^S(v_*) = \min\{\Psi_i(\delta') : \delta \in S(v_*)\}$ for $i = 1, \ldots, m$, i.e., δ is optimal in $S(v_*)$ relative to $\Psi_1, \ldots, \Psi_m$ simultaneously.

Assume that $\Psi_1, \ldots, \Psi_{m-1}$ are strictly increasing cost functions and Ψ_m is increasing or strictly increasing. We now describe how to recognize the existence of an element from $S(v_*)$, which is totally optimal in $S(v_*)$ relative to $\Psi_1, \ldots, \Psi_m$.

For $i = 1, \ldots, m$, we apply the algorithm $\mathscr{A}_1$ to the circuit S and the cost function Ψ_i. As a result, we obtain for $i = 1, \ldots, m$ the edge-preserving subcircuit S^{Ψ_i} of the circuit S and the number $\Psi_i^S(v_*)$ assigned to the node v_* of the circuit S^{Ψ_i}. Next, we apply to S sequentially the procedure of optimization (the algorithm $\mathscr{A}_1$) relative to the cost functions $\Psi_1, \ldots, \Psi_m$. As a result, we obtain circuits $S^{\Psi_1}, S^{\Psi_1,\Psi_2}, \ldots, S^{\Psi_1,\ldots,\Psi_m}$ and numbers $\varphi_1, \varphi_2, \ldots, \varphi_m$ assigned to the node v_* of these circuits. From Theorem 3.1 it follows that $\varphi_1 = \min\{\Psi_1(\delta) : \delta \in S(v_*)\}$ and $\varphi_i = \min\{\Psi_i(\delta) : \delta \in S^{\Psi_1,\ldots,\Psi_{i-1}}(v_*)\}$ for $i = 2, \ldots, m$. One can show that an element from $S(v_*)$, which is totally optimal in $S(v_*)$ relative to $\Psi_1, \ldots, \Psi_m$ exists if and only if $\Psi_i^S(v_*) = \varphi_i$ for $i = 1, \ldots, m$.

3.3 Algorithm for Counting Elements

Let S be a circuit without repetitions. We now describe an algorithm $\mathscr{A}_2$, which assigns to each node v of S the number $C(v) = |S(v)|$.

Algorithm $\mathscr{A}_2$ (counting elements)
Input: A circuit S without repetitions.
Output: The circuit S with a number $C(v)$ assigned to each node v of S.

1. If all nodes of S are processed, then return the obtained circuit.
2. Otherwise, choose a node v of S that is not processed yet and which is either an input node or a node for which all parents are processed.
3. If v is an input node, then assign to it the number $C(v) = 1$, label it as processed, and proceed to step 1.
4. If v is a functional node with $k = k(v)$ entering edges leaving the nodes $v_1, \ldots, v_k$, respectively, then assign to v the number $C(v) = C(v_1) \cdot \ldots \cdot C(v_k)$, label v as processed, and proceed to step 1.
5. If v is a unifying node with m entering edges leaving the nodes $v_1, \ldots, v_m$, respectively, then assign to v the number $C(v) = C(v_1) + \cdots + C(v_m)$, label v as processed, and proceed to step 1.

Remark 3.3 Note that the algorithm $\mathscr{A}_2$ can work with the schemes of circuits instead of the circuits (see Sect. 2.4).

Remark 3.4 We say that a circuit S is reachable if, for any node v of S, there is a directed path from v to the output v_* of S. Let S be a reachable circuit without repetitions and T be an edge-preserving subcircuit of S. Then the number $|S(v_*)|$ is an upper bound on the absolute values of the numbers at the inputs of operations (additions and multiplications) of the algorithm $\mathscr{A}_2$ processing T. Assume the contrary. In this case, it is easy to show that $C(v_*) > |S(v_*)|$, which contradicts Theorem 3.2.

Remark 3.5 Later we will use the following way to transform a circuit into a reachable one. Let S be a circuit and S' be a circuit obtained from S by the removal of all nodes such that there is no directed path from the considered node to the output. We also remove all edges that are incident to the removed nodes. One can show that, for any node v of S', $S'(v) = S(v)$. In particular, $S'(v_*) = S(v_*)$.

Theorem 3.2 *Let S be a circuit without repetitions. Then, for any node v of S, $C(v) = |S(v)|$. For any input node of S, the algorithm $\mathscr{A}_2$ does not make any operations. For any functional node of S with k entering edges, the algorithm $\mathscr{A}_2$ makes $k - 1$ multiplications. For any unifying node of S with m entering edges, the algorithm $\mathscr{A}_2$ makes $m - 1$ additions.*

Proof We prove by induction on nodes of S that $C(v) = |S(v)|$ for any node v of S. Let v be an input node. Then $S(v) = \{a_v\}$ and, evidently, $C(v) = 1 = |S(v)|$. To compute the value $C(v) = 1$, the algorithm $\mathscr{A}_2$ does not make any operations.

Let v be a functional node with $k = k(v)$ entering edges $e_1, \ldots, e_k$, which are labeled with numbers $1, \ldots, k$ and leave the nodes $v_1, \ldots, v_k$ such that $C(v_1) = |S(v_1)|, \ldots, C(v_k) = |S(v_k)|$. By definition,

$$S(v) = \{f_v(\delta_1, \ldots, \delta_k) : \delta_1 \in S(v_1), \ldots, \delta_k \in S(v_k)\} .$$

Let $\delta_1', \delta_1'' \in S(v_1), \ldots, \delta_k', \delta_k'' \in S(v_k)$ and $(\delta_1', \ldots, \delta_k') \neq (\delta_1'', \ldots, \delta_k'')$. Since f_v is injective, the elements $f_v(\delta_1', \ldots, \delta_k')$ and $f_v(\delta_1'', \ldots, \delta_k'')$ are different. Therefore, $|S(v)| = |S(v_1)| \cdot \ldots \cdot |S(v_k)|$. Using the inductive hypothesis, we obtain that $C(v) = |S(v)|$. To compute the value $C(v) = C(v_1) \cdot \ldots \cdot C(v_k)$, the algorithm $\mathscr{A}_2$ makes $k - 1$ multiplications.

Let v be a unifying node with m entering edges $e_1, \ldots, e_m$, which leave the nodes $v_1, \ldots, v_m$ such that $C(v_1) = |S(v_1)|, \ldots, C(v_m) = |S(v_m)|$. By definition, $S(v) = S(v_1) \cup \cdots \cup S(v_m)$. Since S is a circuit without repetitions, the sets $S(v_1), \ldots, S(v_m)$ are pairwise disjoint. Therefore, $|S(v)| = |S(v_1)| + \cdots + |S(v_m)|$. Using the inductive hypothesis, we obtain that $C(v) = |S(v)|$. To compute the value $C(v) = C(v_1) + \cdots + C(v_m)$, the algorithm $\mathscr{A}_2$ makes $m - 1$ additions. □

References

1. AbouEisha, H., Amin, T., Chikalov, I., Hussain, S., Moshkov, M.: Extensions of Dynamic Programming for Combinatorial Optimization and Data Mining, Intelligent Systems Reference Library, vol. 146. Springer (2019)
2. Amin, T., Moshkov, M.: Totally optimal decision rules. Discret. Appl. Math. **236**, 453–458 (2018)
3. Azad, M., Moshkov, M.: Multi-stage optimization of decision and inhibitory trees for decision tables with many-valued decisions. Eur. J. Oper. Res. **263**(3), 910–921 (2017)
4. Chikalov, I., Hussain, S., Moshkov, M.: Totally optimal decision trees for Boolean functions. Discret. Appl. Math. **215**, 1–13 (2016)
5. Mankowski, M., Moshkov, M.: Extensions of dynamic programming for multi-stage combinatorial optimization. Theor. Comput. Sci. **844**, 106–132 (2020)

Chapter 4
Bi-Criteria Optimization of Elements

In this chapter, we take special attention to bi-criteria optimization problems relative to two increasing cost functions, and propose an algorithm $\mathscr{A}_4$ for the construction of the set of Pareto optimal points for such a problem.

First, we consider some tools for the study of Pareto optimal points. After that, we design and study the algorithm $\mathscr{A}_4$. Next, we discuss some possibility of decreasing the number of operations made by this algorithm in the case when functional nodes in the circuit can have many entering edges. In the end, we show how the set of Pareto optimal points can be transformed into the graphs of functions that describe the relationships between the considered cost functions.

We can use the algorithm $\mathscr{A}_4$ to recognize the existence of an element, which is optimal relative to two increasing cost functions at the same time (a totally optimal element). Such an element exists if and only if there is only one Pareto optimal point for bi-criteria optimization problem relative to the considered cost functions.

Algorithms similar in some sense to the algorithm $\mathscr{A}_4$ were considered in [1, 3] for bi-criteria optimization of decision trees, decision rules, decision rule systems, and element partition trees. These problems were studied separately with their own algorithms and proofs. The first two sections of this chapter are based on [5].

4.1 Tools for Study of Pareto Optimal Points

To study a bi-criteria optimization problem, we correspond to each element a point, which coordinates are values of the two considered cost functions for the given element. We aim to construct for this problem all Pareto optimal points (nondominated points). It is easy to do if the number of elements under consideration is reasonable. However, if the number of elements can be huge, we should use special algorithms, which are based on the tools discussed in this section. For the sake of completeness, we provide proofs of some statements considered in [1].

M. Mankowski and M. Moshkov, *Dynamic Programming Multi-Objective Combinatorial Optimization*, Studies in Systems, Decision and Control 331,
https://doi.org/10.1007/978-3-030-63920-4_4

Let $\mathbb{R}^2$ be the set of pairs of real numbers (points). We consider a partial order $\le$ on the set $\mathbb{R}^2$ (on the plane): $(c, d) \le (a, b)$ if $c \le a$ and $d \le b$. Two points α and β are comparable if $\alpha \le \beta$ or $\beta \le \alpha$. A subset of $\mathbb{R}^2$ in which no two different points are comparable is called an antichain. We will write $\alpha < \beta$ if $\alpha \le \beta$ and $\alpha \ne \beta$. If α and β are comparable, then $\min(\alpha, \beta) = \alpha$ if $\alpha \le \beta$ and $\min(\alpha, \beta) = \beta$ if $\alpha > \beta$.

Let A be a nonempty finite subset of $\mathbb{R}^2$. A point $\alpha \in A$ is called a Pareto optimal point (POP) for A if there is no a point $\beta \in A$ such that $\beta < \alpha$. We denote by $Par(A)$ the set of Pareto optimal points for A. It is clear that $Par(A)$ is an antichain.

Lemma 4.1 [1] *Let A be a nonempty finite subset of the set $\mathbb{R}^2$. Then, for any point $\alpha \in A$, there is a point $\beta \in Par(A)$ such that $\beta \le \alpha$.*

Proof Let $\beta = (a, b)$ be a point from A such that $(a, b) \le \alpha$ and $a + b = \min\{c + d : (c, d) \in A, (c, d) \le \alpha\}$. Then $(a, b) \in Par(A)$. □

Lemma 4.2 [1] *Let A be a nonempty finite subset of $\mathbb{R}^2$, $B \subseteq A$, and $Par(A) \subseteq B$. Then $Par(B) = Par(A)$.*

Proof It is clear that $Par(A) \subseteq Par(B)$. Let us assume that, for some β, $\beta \in Par(B)$ and $\beta \notin Par(A)$. Then there exists $\alpha \in A$ such that $\alpha < \beta$. By Lemma 4.1, there exists $\gamma \in Par(A) \subseteq B$ such that $\gamma \le \alpha$. Therefore, $\gamma < \beta$ and $\beta \notin Par(B)$. Hence, $Par(B) = Par(A)$. □

Lemma 4.3 [1] *Let $A_1, \ldots, A_k$ be nonempty finite subsets of $\mathbb{R}^2$. Then $Par(A_1 \cup \cdots \cup A_k) \subseteq Par(A_1) \cup \cdots \cup Par(A_k)$.*

Proof Let $\alpha \in (A_1 \cup \cdots \cup A_k) \setminus (Par(A_1) \cup \cdots \cup Par(A_k))$. Then there is $i \in \{1, \ldots, k\}$ such that $\alpha \in A_i$ but $\alpha \notin Par(A_i)$. Therefore, there is $\beta \in A_i$ such that $\beta < \alpha$. Hence, $\alpha \notin Par(A_1 \cup \cdots \cup A_k)$, and $Par(A_1 \cup \cdots \cup A_k) \subseteq Par(A_1) \cup \cdots \cup Par(A_k)$. □

Proposition 4.1 *Let $A_1, \ldots, A_k$ be nonempty finite subsets of $\mathbb{R}^2$. Then $Par(A_1 \cup \cdots \cup A_k) = Par(Par(A_1) \cup \cdots \cup Par(A_k))$.*

Proof The considered statement follows from Lemmas 4.3 and 4.2. □

Let f, g be increasing functions from $\mathbb{R}^k$ to $\mathbb{R}$, and $A_1, \ldots, A_k$ be nonempty finite subsets of the set $\mathbb{R}^2$. We denote by $J_{f,g}(A_1, \ldots, A_k)$ the set

$$\{(f(a_1, \ldots, a_k), g(b_1, \ldots, b_k)) : (a_1, b_1) \in A_1, \ldots, (a_k, b_k) \in A_k\} .$$

Lemma 4.4 *Let $A_1, \ldots, A_k$ be nonempty finite subsets of $\mathbb{R}^2$, and f, g be increasing functions from $\mathbb{R}^k$ to $\mathbb{R}$. Then $Par(J_{f,g}(A_1, \ldots, A_k)) \subseteq J_{f,g}(Par(A_1), \ldots, Par(A_k))$.*

Proof Let $\beta \in Par(J_{f,g}(A_1, \ldots, A_k))$ and $\beta = (f(a_1, \ldots, a_k), g(b_1, \ldots, b_k))$, where $(a_1, b_1) \in A_1, \ldots, (a_k, b_k) \in A_k$. Using Lemma 4.1, we obtain that, for $i = 1, \ldots, k$, there exists $(a_i', b_i') \in Par(A_i)$ such that $(a_i', b_i') \le (a_i, b_i)$. Since f and

g are increasing functions, $\alpha = (f(a'_1, \ldots, a'_k), g(b'_1, \ldots, b'_k)) \leq (f(a_1, \ldots, a_k), g(b_1, \ldots, b_k)) = \beta$. It is clear that $\alpha \in J_{f,g}(Par(A_1), \ldots, Par(A_k))$. Taking into account that

$$\beta \in Par(J_{f,g}(A_1, \ldots, A_k)) ,$$

we obtain $\beta = \alpha$. Therefore, $Par(J_{f,g}(A_1, \ldots, A_k)) \subseteq J_{f,g}(Par(A_1), \ldots, Par(A_k))$. □

Proposition 4.2 *Let $A_1, \ldots, A_k$ be nonempty finite subsets of $\mathbb{R}^2$, and f, g be increasing functions from $\mathbb{R}^k$ to $\mathbb{R}$. Then*

$$Par(J_{f,g}(A_1, \ldots, A_k)) = Par(J_{f,g}(Par(A_1), \ldots, Par(A_k))) .$$

Proof The considered statement follows from Lemmas 4.4 and 4.2. □

Points from $Par(A)$ can be ordered in the following way: $(a_1, b_1), \ldots, (a_t, b_t)$, where $a_1 < \cdots < a_t$. Since points from $Par(A)$ are incomparable, $b_1 > \cdots > b_t$. We will refer to the sequence $(a_1, b_1), \ldots, (a_t, b_t)$ as the normal representation of the set $Par(A)$.

It is well known (see [4]) that there exists an algorithm, which for a given set A, constructs the set $Par(A)$ and makes $O(|A| \log |A|)$ comparisons.

For the sake of completeness, we describe an algorithm, which for a given nonempty finite subset A of the set $\mathbb{R}^2$, constructs the normal representation of the set $Par(A)$. We assume that A is a multiset containing, possibly, repeating elements. The cardinality $|A|$ of A is the total number of elements in A.

Algorithm $\mathscr{A}_3$ (construction of normal representation for the set of POPs).

Input: A nonempty finite subset A of the set $\mathbb{R}^2$ containing, possibly, repeating elements (multiset).

Output: Normal representation P of the set $Par(A)$ of Pareto optimal points for A.

1. Set P equal to the empty sequence.
2. Using the merge sort (see [2]), construct a sequence B of all points from A ordered according to the first coordinate in the ascending order.
3. If there is only one point in the sequence B, then add this point to the end of the sequence P, return P, and finish the work of the algorithm. Otherwise, choose the first $\alpha = (\alpha_1, \alpha_2)$ and the second $\beta = (\beta_1, \beta_2)$ points from B.
4. If α and β are comparable, then remove α and β from B, add the point $\min(\alpha, \beta)$ to the beginning of B, and proceed to step 3.
5. If α and β are not comparable (in this case $\alpha_1 < \beta_1$ and $\alpha_2 > \beta_2$), then remove α from B, add the point α to the end of P, and proceed to step 3.

Proposition 4.3 *Let A be a nonempty finite subset of the set $\mathbb{R}^2$ containing, possibly, repeating elements (multiset). Then the algorithm $\mathscr{A}_3$ returns the normal representation of the set $Par(A)$ of Pareto optimal points for A and makes at most $4\,|A| \log_2 |A|$ comparisons.*

Proof In step 2, we use the merge sort, which requires at most $2|A|\log_2|A|+1$ comparisons (see [2]). Each call to step 3 (with the exception of the last one, where the algorithm does not make comparisons) leads to two comparisons. The number of calls to step 3 is equal to $|A|$. Therefore, the algorithm $\mathscr{A}_3$ makes at most $2|A|\log_2|A| + 1 + 2(|A|-1) < 2|A|\log_2|A| + 2|A|$ comparisons. If $|A|=1$, then the algorithm $\mathscr{A}_3$ makes $0 = 4|A|\log_2|A|$ comparisons. If $|A| \geq 2$, then $2|A|\log_2|A| + 2|A| \leq 4|A|\log_2|A|$.

Let the output sequence P be equal to $(a_1, b_1), \ldots, (a_t, b_t)$ and let us set $Q = \{(a_1, b_1), \ldots, (a_t, b_t)\}$. It is clear that $a_1 < \cdots < a_t, b_1 > \cdots > b_t$ and, for any $\alpha \in A, \alpha \notin Q$, there exists $\beta \in Q$ such that $\beta < \alpha$. From here it follows that $Par(A) \subseteq Q$ and Q is an antichain. Let us assume that there exists $\gamma \in Q$, which does not belong to $Par(A)$. Then there exists $\alpha \in A$ such that $\alpha < \gamma$. Since Q is an antichain, $\alpha \notin Q$. In this case, there exists $\beta \in Q$ such that $\beta < \alpha$. This results in two different points β and γ from Q being comparable, which is impossible. Therefore, $Q = Par(A)$ and P is the normal representation of the set $Par(A)$. □

Lemma 4.5 [1] *Let A be a nonempty finite subset of $\mathbb{R}^2$. Then*

$$|Par(A)| \leq \min\left(\left|A^{(1)}\right|, \left|A^{(2)}\right|\right),$$

where $A^{(1)} = \{a : (a, b) \in A\}$ and $A^{(2)} = \{b : (a, b) \in A\}$.

Proof Let $(a, b), (c, d) \in Par(A)$ and $(a, b) \neq (c, d)$. Then $a \neq c$ and $b \neq d$ (otherwise, (a, b) and (c, d) are comparable). Therefore, $|Par(A)| \leq \min\left(\left|A^{(1)}\right|, \left|A^{(2)}\right|\right)$. □

Let A be a nonempty finite subset of $\mathbb{R}^2$. We correspond to A a partial function $\mathscr{F}_A : \mathbb{R} \to \mathbb{R}$ defined in the following way: $\mathscr{F}_A(x) = \min\{b : (a, b) \in A, a \leq x\}$ for any $x \in \mathbb{R}$.

Lemma 4.6 [1] *Let A be a nonempty finite subset of $\mathbb{R}^2$, and $(a_1, b_1), \ldots, (a_t, b_t)$ be the normal representation of the set $Par(A)$. Then, for any $x \in \mathbb{R}$, $\mathscr{F}_A(x) = \mathscr{F}(x)$, where*

$$\mathscr{F}(x) = \begin{cases} undefined, & x < a_1 \\ b_1, & a_1 \leq x < a_2 \\ \ldots & \ldots \\ b_{t-1}, & a_{t-1} \leq x < a_t \\ b_t, & a_t \leq x \end{cases}.$$

Proof One can show that $a_1 = \min\{a : (a, b) \in A\}$. Therefore, the value $\mathscr{F}_A(x)$ is undefined if $x < a_1$. Let $x \geq a_1$. Then both values $\mathscr{F}(x)$ and $\mathscr{F}_A(x)$ are defined. It is easy to check that $\mathscr{F}(x) = \mathscr{F}_{Par(A)}(x)$. Since $Par(A) \subseteq A$, we have $\mathscr{F}_A(x) \leq \mathscr{F}(x)$. By Lemma 4.1, for any point $(a, b) \in A$, there is a point $(a_i, b_i) \in Par(A)$ such that $(a_i, b_i) \leq (a, b)$. Therefore, $\mathscr{F}(x) \leq \mathscr{F}_A(x)$ and $\mathscr{F}_A(x) = \mathscr{F}(x)$. □

Remark 4.1 [1] Let A be a nonempty finite subset of $\mathbb{R}^2$, $(a_1, b_1), \ldots, (a_t, b_t)$ be the normal representation of the set $Par(A)$, and $rev(A) = \{(b, a) : (a, b) \in A\}$. Then

$Par(rev(A)) = rev(Par(A))$ and $(b_t, a_t), \ldots, (b_1, a_1)$ is the normal representation of the set $Par(rev(A))$.

Remark 4.2 [1] We can consider not only function $\mathscr{F}_A$ but also function $\mathscr{F}_{rev(A)} : \mathbb{R} \to \mathbb{R}$ defined in the following way:

$$\mathscr{F}_{rev(A)}(x) = \min\{a : (b, a) \in rev(A), b \leq x\} = \min\{a : (a, b) \in A, b \leq x\} .$$

From Remark 4.1 and Lemma 4.6 it follows that

$$\mathscr{F}_{rev(A)}(x) = \begin{cases} undefined, & x < b_t \\ a_t, & b_t \leq x < b_{t-1} \\ \ldots & \ldots \\ a_2, & b_2 \leq x < b_1 \\ a_1, & b_1 \leq x \end{cases} .$$

4.2 Bi-Criteria Optimization of Elements

Let S be a circuit without repetitions and Ψ, Φ be increasing cost functions for S given, respectively, by the functions ψ, ψ_v and φ, φ_v, $v \in V_f(S)$. For each node v of the circuit S, we denote $t_{\Psi,\Phi}(S, v) = \{(\Psi(\delta), \Phi(\delta)) : \delta \in S(v)\}$.

It is easy to describe the set $t_{\Psi,\Phi}(S, v)$ for each node v of S in an inductive way:

- If v is an input node of S, then $t_{\Psi,\Phi}(S, v) = \{(\psi(a_v), \varphi(a_v))\}$.
- If v is a functional node of S with $k = k(v)$ entering edges $e_1, \ldots, e_k$, which are labeled with numbers $1, \ldots, k$ and leave nodes $v_1, \ldots, v_k$, respectively, then

$$\begin{aligned} t_{\Psi,\Phi}(S, v) &= J_{\psi_v,\varphi_v}(t_{\Psi,\Phi}(S, v_1), \ldots, t_{\Psi,\Phi}(S, v_k)) \\ &= \{(\psi_v(a_1, \ldots, a_k), \varphi_v(b_1, \ldots, b_k)) \\ &\quad : (a_1, b_1) \in t_{\Psi,\Phi}(S, v_1), \ldots, (a_k, b_k) \in t_{\Psi,\Phi}(S, v_k)\} . \end{aligned}$$

- If v is a unifying node of S with m entering edges $e_1, \ldots, e_m$, which leave nodes $v_1, \ldots, v_m$, respectively, then $t_{\Psi,\Phi}(S, v) = t_{\Psi,\Phi}(S, v_1) \cup \cdots \cup t_{\Psi,\Phi}(S, v_m)$.

We denote $Par(t_{\Psi,\Phi}(S, v))$ the set of Pareto optimal points for the set $t_{\Psi,\Phi}(S, v)$. We now describe an algorithm $\mathscr{A}_4$, which constructs the set $Par(t_{\Psi,\Phi}(S, v_*))$, where v_* is the output of S. In fact, this algorithm constructs, for each node v of the circuit S, the set $B(v) = Par(t_{\Psi,\Phi}(S, v))$.

Algorithm $\mathscr{A}_4$ (bi-criteria optimization of elements).

Input: Circuit S without repetitions and increasing cost functions Ψ, Φ for S given, respectively, by the functions ψ, ψ_v and φ, φ_v, $v \in V_f(S)$.

Output: The set $Par(t_{\Psi,\Phi}(S, v_*))$ of Pareto optimal points for the set of pairs $t_{\Psi,\Phi}(S, v_*) = \{(\Psi(\delta), \Phi(\delta)) : \delta \in S(v_*)\}$, where v_* is the output of S.

1. If all nodes of S are processed, then return the set $B(v_*)$. Otherwise, choose a node v in the circuit S, which is not processed yet and which is either an input node of S or a node v of S for which all parents are already processed, i.e., the set $B(v_i)$ is already constructed for each parent v_i of the node v.
2. If v is an input node, then set $B(v) = \{(\psi(a_v), \varphi(a_v))\}$. Mark the node v as processed and proceed to step 1.
3. If v is a functional node with $k = k(v)$ entering edges, which are labeled with numbers $1, \ldots, k$ and leave nodes $v_1, \ldots, v_k$, respectively, then construct the multiset $E(v) = \{(\psi_v(a_1, \ldots, a_k), \varphi_v(b_1, \ldots, b_k)) : (a_1, b_1) \in B(v_1), \ldots, (a_k, b_k) \in B(v_k)\}$ by computation of values $\psi_v(a_1, \ldots, a_k)$ and $\varphi_v(b_1, \ldots, b_k)$ for all pairs $(a_1, b_1) \in B(v_1), \ldots, (a_k, b_k) \in B(v_k)$. Apply to $E(v)$ the algorithm $\mathscr{A}_3$, which constructs the set $Par(E(v))$. Set $B(v) = Par(E(v))$. Mark the node v as processed and proceed to step 1.
4. If v is a unifying node with m entering edges $e_1, \ldots, e_m$ leaving nodes $v_1, \ldots, v_m$, respectively, then construct the multiset $D(v) = B(v_1) \cup \cdots \cup B(v_m)$ by transcription of pairs from the sets $B(v_1), \ldots, B(v_m)$. Apply to $D(v)$ the algorithm $\mathscr{A}_3$, which constructs the set $Par(D(v))$. Set $B(v) = Par(D(v))$. Mark the node v as processed and proceed to step 1.

Remark 4.3 Note that the algorithm $\mathscr{A}_4$ can work with the schemes of circuits instead of the circuits and with adapted cost functions (see Sect. 2.4).

Remark 4.4 Let S be a circuit without repetitions and Ψ, Φ be increasing integer cost functions for S given, respectively, by the functions ψ, ψ_v and φ, φ_v, $v \in V_f(S)$. Denote $N_\Psi(S) = \max\{|\Psi(\delta)| : \delta \in U(S)\}$ and $N_\Phi(S) = \max\{|\Phi(\delta)| : \delta \in U(S)\}$, where $U(S) = \bigcup_{v \in V(S)} S(v)$. Simple analysis of the algorithm $\mathscr{A}_4$ shows that the number $\max(N_\Psi(S), N_\Phi(S))$ is an upper bound on the absolute values of the numbers at the inputs of operations ψ_v and φ_v related to the functional nodes and at the inputs of comparisons related to the functional and unifying nodes of S. When a function ψ_v or a function φ_v is given by a formula containing more than one elementary operation, we need to evaluate the absolute values of the numbers at the inputs of each elementary operation. For all cases studied in this book, it is easy to show that these absolute values are also at most $\max(N_\Psi(S), N_\Phi(S))$ or are at most some upper bound on $\max(N_\Psi(S), N_\Phi(S))$ considered in the corresponding chapter. In particular, let the values of the function ψ be nonnegative and functions ψ_v be given by formulas containing only operations $x + y$, $\max(x, y)$ and nonnegative numeric parameters. Let us assume that the absolute value of at least one number at an input of an elementary operation from the formula for ψ_v is greater than $\max(N_\Psi(S), N_\Phi(S))$. Then the absolute value of this formula is greater than $N_\Psi(S)$, which is impossible.

The same situation is in the case when values of the function ψ are nonpositive and functions ψ_v are given by formulas containing only operations $x + y$, $\min(x, y)$ and nonpositive numeric parameters. These considerations can be extended to the functions φ_v. The operations related to the input nodes of S should be considered separately and only in the cases when values of the function ψ or the function φ are represented not by constants, but by formulas.

Denote $M_{\Psi,\Phi}(S) = \max\{|Par(t_{\Psi,\Phi}(S, v))| : v \in V(S)\}$.

Theorem 4.1 *Let S be a circuit without repetitions and Ψ, Φ be increasing cost functions for S given, respectively, by the functions ψ, ψ_v and φ, φ_v, $v \in V_f(S)$. Then, for each node v of the circuit S, the algorithm $\mathscr{A}_4$ constructs the set $B(v) = Par(t_{\Psi,\Phi}(S, v))$. For any input node of S, the algorithm $\mathscr{A}_4$ computes the values of functions ψ and φ one time. For any functional node v of S with k entering edges, the algorithm $\mathscr{A}_4$ computes the values of functions ψ_v and φ_v at most $M_{\Psi,\Phi}(S)^k$ times and makes at most $4kM_{\Psi,\Phi}(S)^k \log_2 M_{\Psi,\Phi}(S)$ comparisons. For any unifying node of S with m entering edges, the algorithm $\mathscr{A}_4$ makes at most $4mM_{\Psi,\Phi}(S) \log_2(mM_{\Psi,\Phi}(S))$ comparisons.*

Proof We prove by induction on nodes of S that $B(v) = Par(t_{\Psi,\Phi}(S, v))$ for any node v of S. Let v be an input node. Then $S(v) = \{a_v\}$ and, evidently,

$$Par(t_{\Psi,\Phi}(S, v)) = t_{\Psi,\Phi}(S, v) = \{(\psi(a_v), \varphi(a_v))\} = B(v) .$$

To construct the set $B(v)$, the algorithm $\mathscr{A}_4$ computes the values of functions ψ and φ one time.

Let v be a functional node with $k = k(v)$ entering edges $e_1, \ldots, e_k$, which are labeled with numbers $1, \ldots, k$ and leave the nodes $v_1, \ldots, v_k$ such that $B(v_1) = Par(t_{\Psi,\Phi}(S, v_1)), \ldots, B(v_k) = Par(t_{\Psi,\Phi}(S, v_k))$. From the description of the algorithm $\mathscr{A}_4$ it follows that $B(v) = Par(J_{\psi_v,\varphi_v}(B(v_1), \ldots, B(v_k)))$. By the inductive hypothesis, $B(v) = Par(J_{\psi_v,\varphi_v}(Par(t_{\Psi,\Phi}(S, v_1)), \ldots, Par(t_{\Psi,\Phi}(S, v_k))))$. Using Proposition 4.2 we obtain

$$\begin{aligned} &Par(J_{\psi_v,\varphi_v}(Par(t_{\Psi,\Phi}(S, v_1)), \ldots, Par(t_{\Psi,\Phi}(S, v_k)))) \\ &= Par(J_{\psi_v,\varphi_v}(t_{\Psi,\Phi}(S, v_1), \ldots, t_{\Psi,\Phi}(S, v_k))) . \end{aligned}$$

Since $t_{\Psi,\Phi}(S, v) = J_{\psi_v,\varphi_v}(t_{\Psi,\Phi}(S, v_1), \ldots, t_{\Psi,\Phi}(S, v_k))$, $B(v) = Par(t_{\Psi,\Phi}(S, v))$. It is clear that $|B(v_i)| = |Par(t_{\Psi,\Phi}(S, v_i))| \leq M_{\Psi,\Phi}(S)$ for $i = 1, \ldots, k$ and $|E(v)| \leq M_{\Psi,\Phi}(S)^k$. To construct the set $E(v) = \{(\psi_v(a_1, \ldots, a_k), \ \varphi_v(b_1, \ldots, b_k)) : (a_1, b_1) \in B(v_1), \ldots, (a_k, b_k) \in B(v_k)\}$, the algorithm $\mathscr{A}_4$ computes the values of functions ψ_v and φ_v at most $M_{\Psi,\Phi}(S)^k$ times. From Proposition 4.3 it follows that, to construct the set $Par(E(v))$, the algorithm $\mathscr{A}_4$ makes at most $4kM_{\Psi,\Phi}(S)^k \log_2 M_{\Psi,\Phi}(S)$ comparisons.

Let v be a unifying node with m entering edges $e_1, \ldots, e_m$, which leave the nodes $v_1, \ldots, v_m$ such that $B(v_1) = Par(t_{\Psi,\Phi}(S, v_1)), \ldots, B(v_m) = Par(t_{\Psi,\Phi}(S, v_m))$.

From the description of the algorithm $\mathscr{A}_4$ it follows that $B(v) = Par(B(v_1) \cup \cdots \cup B(v_m))$. Using the inductive hypothesis, we obtain that

$$B(v) = Par(Par(t_{\Psi,\Phi}(S, v_1)) \cup \cdots \cup Par(t_{\Psi,\Phi}(S, v_m))) .$$

By Proposition 4.1,

$$\begin{aligned} &Par(Par(t_{\Psi,\Phi}(S, v_1)) \cup \cdots \cup Par(t_{\Psi,\Phi}(S, v_m))) \\ &= Par(t_{\Psi,\Phi}(S, v_1) \cup \cdots \cup t_{\Psi,\Phi}(S, v_m)) . \end{aligned}$$

Since $t_{\Psi,\Phi}(S, v) = t_{\Psi,\Phi}(S, v_1) \cup \cdots \cup t_{\Psi,\Phi}(S, v_m)$, $B(v) = Par(t_{\Psi,\Phi}(S, v))$. It is clear that $|B(v_i)| = \left|Par(t_{\Psi,\Phi}(S, v_i))\right| \le M_{\Psi,\Phi}(S)$ for $i = 1, \ldots, m$ and $|D(v)| \le m M_{\Psi,\Phi}(S)$. From Proposition 4.3 it follows that, to construct the set $Par(D(v))$, the algorithm $\mathscr{A}_4$ makes at most $4m M_{\Psi,\Phi}(S) \log_2(m M_{\Psi,\Phi}(S))$ comparisons. □

We now describe a way to evaluate the parameter

$$M_{\Psi,\Phi}(S) = \max\{\left|Par(t_{\Psi,\Phi}(S, v))\right| : v \in V(S)\} ,$$

where $t_{\Psi,\Phi}(S, v) = \{(\Psi(\delta), \Phi(\delta)) : \delta \in S(v)\}$. Let $N_\Psi(S) = \max\{|\Psi(\delta)| : \delta \in U(S)\}$, where $U(S) = \bigcup_{v \in V(S)} S(v)$.

Proposition 4.4 *Let S be a circuit without repetitions and Ψ, Φ be cost functions for S such that Ψ is either nonnegative integer or nonpositive integer cost function. Then $M_{\Psi,\Phi}(S) \le N_\Psi(S) + 1$.*

Proof Let $v \in V(S)$ and $t_{\Psi,\Phi}(S, v)^{(1)} = \{a : (a, b) \in t_{\Psi,\Phi}(S, v)\} = \{\Psi(\delta) : \delta \in S(v)\}$. Then either $t_{\Psi,\Phi}(S, v)^{(1)} \subseteq \{0, 1, \ldots, N_\Psi(S)\}$ or

$$t_{\Psi,\Phi}(S, v)^{(1)} \subseteq \{-N_\Psi(S), \ldots, -1, 0\} .$$

Therefore, $|t_{\Psi,\Phi}(S, v)^{(1)}| \le N_\Psi(S) + 1$. Using Lemma 4.5 we obtain that

$$\left|Par(t_{\Psi,\Phi}(S, v))\right| \le N_\Psi(S) + 1 .$$

Thus, $M_{\Psi,\Phi}(S) \le N_\Psi(S) + 1$. □

The following more general bound is used for the study of a number of problems.

Proposition 4.5 *Let S be a circuit without repetitions and Ψ, Φ be cost functions for S. Then $M_{\Psi,\Phi}(S) \le |\{\Psi(\delta) : \delta \in U(S)\}|$.*

Proof Let $v \in V(S)$ and $t_{\Psi,\Phi}(S,v)^{(1)} = \{a : (a,b) \in t_{\Psi,\Phi}(S,v)\} = \{\Psi(\delta) : \delta \in S(v)\}$. Then $t_{\Psi,\Phi}(S,v)^{(1)} \subseteq \{\Psi(\delta) : \delta \in U(S)\}$. Therefore, $\left|t_{\Psi,\Phi}(S,v)^{(1)}\right| \le |\{\Psi(\delta) : \delta \in U(S)\}|$. Using Lemma 4.5 we obtain that

$$\left|Par(t_{\Psi,\Phi}(S,v))\right| \le |\{\Psi(\delta) : \delta \in U(S)\}| .$$

Thus, $M_{\Psi,\Phi}(S) \le |\{\Psi(\delta) : \delta \in U(S)\}|$. □

4.2.1 *Totally Optimal Elements*

Let Ψ and Φ be cost functions for S, and δ be an element from the set $S(v_*)$. We remind that δ is called totally optimal in $S(v_*)$ relative to Ψ and Φ if δ is optimal in $S(v_*)$ relative to Ψ and Φ simultaneously.

Assume that Ψ and Φ are increasing cost functions. We now describe how to recognize the existence of an element from $S(v_*)$, which is totally optimal in $S(v_*)$ relative to Ψ and Φ. Using the algorithm $\mathscr{A}_4$, we construct the set $Par(t_{\Psi,\Phi}(S,v_*))$ of Pareto optimal points for the set of pairs $t_{\Psi,\Phi}(S,v_*) = \{(\Psi(\delta),\Phi(\delta)) : \delta \in S(v_*)\}$. A totally optimal element exists if and only if the set $Par(t_{\Psi,\Phi}(S,v_*))$ contains only one Pareto optimal point.

4.3 Modification of the Algorithm $\mathscr{A}_4$: Processing of Functional Nodes

Let S be a circuit without repetitions and Ψ, Φ be increasing cost functions for S given, respectively, by the functions ψ, ψ_v and φ, φ_v, $v \in V_f(S)$. We assume that Ψ is either a nonnegative integer cost function (the first case) or a nonpositive integer cost function (the second case). By Proposition 4.4, $M_{\Psi,\Phi}(S) \le N_\Psi(S) + 1$, where $N_\Psi(S) = \max\{|\Psi(\delta)| : \delta \in U(S)\}$. Let v be a functional node of S with $k = k(v) \ge 3$ entering edges, which are labeled with numbers $1,\ldots,k$ and leave nodes $v_1,\ldots,v_k$, respectively. During the processing of the node v, the algorithm $\mathscr{A}_4$, for given sets $Par(t_{\Psi,\Phi}(S,v_1)),\ldots, Par(t_{\Psi,\Phi}(S,v_k))$, constructs the set $Par(t_{\Psi,\Phi}(S,v))$. According to Theorem 4.1, to process the node v, the algorithm $\mathscr{A}_4$ computes the values of functions ψ_v and φ_v at most $M_{\Psi,\Phi}(S)^k \le (N_\Psi(S)+1)^k$ times and makes at most $4kM_{\Psi,\Phi}(S)^k \log_2 M_{\Psi,\Phi}(S) \le 4k(N_\Psi(S)+1)^k \log_2(N_\Psi(S)+1)$ comparisons. In some special cases, we can modify the way in which the algorithm $\mathscr{A}_4$ processes functional nodes and reduce these upper bounds.

Let $f(x_1,x_2)$ be a function from $\mathbb{R}^2$ to $\mathbb{R}$. For $t = 2,3,\ldots$, we define the function $f_t(x_1,\ldots,x_t)$ from $\mathbb{R}^t$ to $\mathbb{R}$ in the following way: $f_2(x_1,x_2) = f(x_1,x_2)$ and $f_t(x_1,\ldots,x_t) = f(f_{t-1}(x_1,\ldots,x_{t-1}),x_t)$ for $t \ge 3$. One can show that, if $f(x_1,x_2)$ is increasing, then $f_t(x_1,\ldots,x_t)$ is increasing for $t = 2,3,\ldots$.

We say that $f(x_1, x_2)$ is a bounded nonnegative integer function if, for any nonnegative integers x_1 and x_2, $f(x_1, x_2)$ is a nonnegative integer and $f(x_1, x_2) \geq \max(x_1, x_2)$. For example, $x_1 + x_2$ and $\max(x_1, x_2)$ are increasing bounded nonnegative integer functions.

We say that $f(x_1, x_2)$ is a bounded nonpositive integer function if, for any nonpositive integers x_1 and x_2, $f(x_1, x_2)$ is a nonpositive integer and $f(x_1, x_2) \leq \min(x_1, x_2)$. For example, $x_1 + x_2$ and $\min(x_1, x_2)$ are increasing bounded nonpositive integer functions.

The first case: Ψ is a nonnegative integer cost function, $f(x_1, x_2)$ and $g(x_1, x_2)$ are increasing functions from $\mathbb{R}^2$ to $\mathbb{R}$, $f(x_1, x_2)$ is a bounded nonnegative integer function, u is a nonnegative integer, and w is a real number.

Let

$$\psi_v(x_1, \ldots, x_k) = f_k(x_1, \ldots, x_k), \quad \varphi_v(x_1, \ldots, x_k) = g_k(x_1, \ldots, x_k) . \tag{4.1}$$

We now describe how the modified algorithm $\mathscr{A}_4$ processes the functional node v. Denote $A_1 = t_{\Psi,\Phi}(S, v_1), \ldots, A_k = t_{\Psi,\Phi}(S, v_k)$, $B_1 = A_1$, and $B_i = J_{f,g}(B_{i-1}, A_i)$ for $i = 2, \ldots, k$. One can show that $B_k = t_{\Psi,\Phi}(S, v)$. Since Ψ is a nonnegative integer cost function, each number from the sets $A_1^{(1)}, \ldots, A_k^{(1)}, B_k^{(1)}$ is a nonnegative integer, which is at most $N_\Psi(S)$. Since $f(x_1, x_2)$ is a bounded nonnegative integer function, each number from the sets $B_2^{(1)}, \ldots, B_{k-1}^{(1)}$ is a nonnegative integer, which is at most $N_\Psi(S)$ (otherwise, the set $B_k^{(1)}$ contains a number greater than $N_\Psi(S)$ but this is impossible). Denote $N = N_\Psi(S) + 1$. As in the proof of Proposition 4.4, we can prove that the cardinality of each of the sets $Par(A_1), \ldots, Par(A_k), Par(B_1), \ldots, Par(B_k)$ is at most N.

We know sets $Par(A_1), \ldots, Par(A_k)$ and need to construct the set $Par(B_k)$. To this end, we construct step by step the sets $Par(B_2), \ldots, Par(B_k)$. Note that we already know the set $Par(B_1) = Par(A_1)$. From Proposition 4.2 it follows that $Par(B_i) = Par(J_{f,g}(Par(B_{i-1}), Par(A_i)))$ for $i = 2, \ldots, k$. Let $i \in \{2, \ldots, k\}$ and the set $Par(B_{i-1})$ be already constructed. We construct the multiset

$$\begin{aligned} &J_{f,g}(Par(B_{i-1}), Par(A_i)) \\ &= \{(f(a_1, a_2), g(b_1, b_2)) : (a_1, b_1) \in Par(B_{i-1}), (a_2, b_2) \in Par(A_i)\} \end{aligned}$$

by computation of values $f(a_1, a_2)$ and $g(b_1, b_2)$ for all pairs $(a_1, b_1) \in Par(B_{i-1})$, $(a_2, b_2) \in Par(A_i)$. To this end, we need to compute the values of functions f and g at most N^2 times. The cardinality of the multiset

$$J_{f,g}(Par(B_{i-1}), Par(A_i))$$

is at most N^2. We apply to this multiset the algorithm $\mathscr{A}_3$, which constructs the set $Par(J_{f,g}(Par(B_{i-1}), Par(A_i))) = Par(B_i)$. From Proposition 4.3 it follows that the algorithm $\mathscr{A}_3$ makes at most $8N^2 \log_2 N$ comparisons. Therefore, to construct the set $Par(B_i)$ from the sets $Par(B_{i-1})$ and $Par(A_i)$, we need to compute the values of functions f and g at most N^2 times and make at most $8N^2 \log_2 N$ comparisons.

As a result, for the functional node v, the modified algorithm $\mathscr{A}_4$ computes the values of functions f and g at most $(k-1)(N_\Psi(S)+1)^2$ times and makes at most $8(k-1)(N_\Psi(S)+1)^2 \log_2(N_\Psi(S)+1)$ comparisons.

Let now

$$\psi_v(x_1, \ldots, x_k) = f_{k+1}(x_1, \ldots, x_k, u), \quad \varphi_v(x_1, \ldots, x_k) = g_{k+1}(x_1, \ldots, x_k, w) . \tag{4.2}$$

We now describe how the modified algorithm $\mathscr{A}_4$ processes the functional node v. Denote $A_1 = t_{\Psi,\Phi}(S, v_1), \ldots, A_k = t_{\Psi,\Phi}(S, v_k)$, $A_{k+1} = \{(u, w)\}$, $B_1 = A_1$, and $B_i = J_{f,g}(B_{i-1}, A_i)$ for $i = 2, \ldots, k+1$. One can show that $B_{k+1} = t_{\Psi,\Phi}(S, v)$.

We know sets $Par(A_1), \ldots, Par(A_k)$, $Par(A_{k+1}) = \{(u, w)\}$, and need to construct the set $Par(B_{k+1})$. To this end, we step by step construct the sets $Par(B_1) = Par(A_1), Par(B_2), \ldots, Par(B_{k+1})$ in the same way as it was described earlier. It is easy to show that the cardinality of each of the sets

$$Par(A_1), \ldots, Par(A_{k+1}), Par(B_1), \ldots, Par(B_{k+1})$$

is at most $N_\Psi(S)+1$. We can prove in the same way as it was done earlier that, to process the functional node v, the modified algorithm $\mathscr{A}_4$ computes the values of functions f and g at most $k(N_\Psi(S)+1)^2$ times and makes at most $8k(N_\Psi(S)+1)^2 \log_2(N_\Psi(S)+1)$ comparisons.

The second case: Ψ is a nonpositive integer cost function, $f(x_1, x_2)$ and $g(x_1, x_2)$ are increasing functions from $\mathbb{R}^2$ to $\mathbb{R}$, $f(x_1, x_2)$ is a bounded nonpositive integer function, u is a nonpositive integer, and w is a real number.

For functions ψ_v and φ_v described by equations (4.1) or (4.2), we can modify the work of the algorithm $\mathscr{A}_4$ in the same way as it was done in the first case. We can prove the following upper bounds on the number of operations made by the modified algorithm $\mathscr{A}_4$ during the processing of the functional node v.

For functions ψ_v and φ_v described by equations (4.1), the modified algorithm $\mathscr{A}_4$ computes the values of functions f and g at most $(k-1)(N_\Psi(S)+1)^2$ times and makes at most $8(k-1)(N_\Psi(S)+1)^2 \log_2(N_\Psi(S)+1)$ comparisons.

For functions ψ_v and φ_v described by equations (4.2), the modified algorithm $\mathscr{A}_4$ computes the values of functions f and g at most $k(N_\Psi(S)+1)^2$ times and makes at most $8k(N_\Psi(S)+1)^2 \log_2(N_\Psi(S)+1)$ comparisons.

Note that the complexity of computation of functions f and g is usually less than the complexity of computation of functions ψ_v and φ_v described by equations (4.1) or (4.2). Note also that the modified algorithm $\mathscr{A}_4$ can work with the schemes of circuits instead of the circuits and with adapted cost functions (see Sect. 2.4). The considered approach is similar to the fusion of sets of POPs described in [1] (see also Sect. 16.1).

Remark 4.5 Let each of the functions Ψ and Φ be either increasing nonnegative or increasing nonpositive integer cost function, $f = f(x_1, x_2)$ and $g = g(x_1, x_2)$ be correspondingly increasing bounded nonnegative or nonpositive integer functions (in particular, if Ψ is nonnegative, then f is nonnegative), u and w be correspondingly nonnegative or nonpositive integers (in particular, if Ψ is nonnegative, then u is nonnegative), v be a functional node of S, and functions ψ_v and φ_v be described by equations (4.1) or (4.2). One can show that the number $\max(N_\Psi(S), N_\Phi(S))$ is an upper bound on the absolute values of the numbers at the inputs of operations (computations of the values of functions f and g, and comparisons) of the modified algorithm $\mathscr{A}_4$ during the processing of the node v.

4.4 Relationships Between Two Cost Functions

Let S be a circuit without repetitions, v_* be the output of S, and Ψ, Φ be increasing cost functions for S. To study relationships between cost functions Ψ and Φ on the set of elements $S(v_*)$, we consider partial functions $\mathscr{F}_S^{\Psi,\Phi} : \mathbb{R} \to \mathbb{R}$ and $\mathscr{F}_S^{\Phi,\Psi} : \mathbb{R} \to \mathbb{R}$ such that, for any $x \in \mathbb{R}$,

$$\mathscr{F}_S^{\Psi,\Phi}(x) = \min\{\Phi(\delta) : \delta \in S(v_*), \Psi(\delta) \leq x\} ,$$

$$\mathscr{F}_S^{\Phi,\Psi}(x) = \min\{\Psi(\delta) : \delta \in S(v_*), \Phi(\delta) \leq x\} .$$

Proposition 4.6 *Let S be a circuit without repetitions, v_* be the output of S, Ψ, Φ be increasing cost functions for S, and $(a_1, b_1), \ldots, (a_k, b_k)$ be the normal representation of the set $Par(t_{\Psi,\Phi}(S, v_*))$, where $a_1 < \cdots < a_k$ and $b_1 > \cdots > b_k$. Then, for any $x \in \mathbb{R}$,*

$$\mathscr{F}_S^{\Psi,\Phi}(x) = \begin{cases} \textit{undefined}, & x < a_1 \\ b_1, & a_1 \leq x < a_2 \\ \ldots & \ldots \\ b_{k-1}, & a_{k-1} \leq x < a_k \\ b_k, & a_k \leq x \end{cases} ,$$

$$\mathscr{F}_S^{\Phi,\Psi}(x) = \begin{cases} \textit{undefined}, & x < b_k \\ a_k, & b_k \leq x < b_{k-1} \\ \ldots & \ldots \\ a_2, & b_2 \leq x < b_1 \\ a_1, & b_1 \leq x \end{cases} .$$

Proof The considered statement follows from Lemma 4.6 and Remark 4.2. □

References

1. AbouEisha, H., Amin, T., Chikalov, I., Hussain, S., Moshkov, M.: Extensions of Dynamic Programming for Combinatorial Optimization and Data Mining, Intelligent Systems Reference Library, vol. 146. Springer (2019)
2. Alekseev, V.B.: Introduction to Algorithm Complexity (in Russian). Moscow State University (2002)
3. Chikalov, I., Hussain, S., Moshkov, M.: Bi-criteria optimization of decision trees with applications to data analysis. Eur. J. Oper. Res. **266**(2), 689–701 (2018)
4. Kung, H.T., Luccio, F., Preparata, F.P.: On finding the maxima of a set of vectors. J. ACM **22**(4), 469–476 (1975)
5. Mankowski, M., Moshkov, M.: Dynamic programming bi-criteria combinatorial optimization. Discret. Appl. Math. **284**, 513–533 (2020)

Part II
Combinatorial Optimization Problems. Conventional Circuits

In this part, we study nine combinatorial optimization problems. For each problem, we construct a conventional circuit without repetitions that represents all elements under consideration (the set of feasible solutions), describe at least two increasing or strictly increasing cost functions for this circuit each of which is adapted (see Sect. 2.4), evaluate the number of operations and the time required by the algorithms $\mathcal{A}_1$, $\mathcal{A}_2$, and $\mathcal{A}_4$, consider an example of the problem instance, and discuss the results of experiments with randomly generated instances of the problem. For the considered problems and cost functions, the algorithms $\mathcal{A}_1$, $\mathcal{A}_2$, and $\mathcal{A}_4$ work with the schemes of circuits instead of the circuits (see Sect. 2.4).

In Chap. 5, we study the problem of matrix chain multiplication.

In Chap. 6, we consider the global sequence alignment problem.

In Chap. 7, we discuss the problem of optimal paths in directed graphs.

In Chap. 8, we consider the binary search tree problem.

In Chap. 9, we study the problem of convex polygon triangulation.

In Chap. 10, we discuss the line breaking (text justification) problem.

In Chap. 11, we study the problem of one-dimensional clustering.

In Chap. 12, we consider the optimal bitonic tour problem.

In Chap. 13, we study the problem of segmented least squares.

Chapter 5
Matrix Chain Multiplication

Matrix chain multiplication is one of the classic optimization problems in computer science. For a given sequence $A_1, A_2, \ldots, A_n$ of matrices, we need to compute the product of these matrices using the minimum number of scalar multiplications on a single processor machine. This problem was introduced by Godbole in [3], where he proposed an $O(n^3)$ dynamic programming algorithm. Hu and Shing presented an $O(n \log n)$ algorithm for this problem in [4, 5]. Other optimization scenarios are also possible (some of the corresponding optimization criteria are discussed in this chapter).

We want to compute the product $A_1 \times A_2 \times \cdots \times A_n$. It is well known that matrix multiplication is associative, i.e., $A \times (B \times C) = (A \times B) \times C$. The cost of multiplying a chain of n matrices depends on the order of multiplications. Each possible ordering of multiplication of n matrices corresponds to a different parenthesization. We consider three cost functions for parenthesizations. An algorithm for multi-stage optimization of parenthesizations relative to a sequence of cost functions was proposed in [1]. In this chapter, we design a conventional circuit without repetitions for the representation of all parenthesizations for a given chain, and evaluate the number of operations and the time required by the algorithms for multi-stage optimization of parenthesizations, for counting optimal parenthesizations, and for bi-criteria optimization of parenthesizations relative to two cost functions. This chapter is partially based on [6].

5.1 Definition of Circuit and Cost Functions

Let $A_1, A_2, \ldots, A_n$, $n \geq 2$, be matrices with sizes $m_0 \times m_1$, $m_1 \times m_2$, ..., $m_{n-1} \times m_n$, respectively. The number n is the dimension of problems under consideration related to the multiplication of these matrices. Binary representations of numbers

M. Mankowski and M. Moshkov, *Dynamic Programming Multi-Objective Combinatorial Optimization*, Studies in Systems, Decision and Control 331,
https://doi.org/10.1007/978-3-030-63920-4_5

$m_0, \ldots, m_n$ form the representation of the problem instance (later, we will omit the word instance). Since n is at most the number of problem parameters, n is at most the length of the problem representation.

We consider not only the whole chain $M(1, n) = A_1 \times \cdots \times A_n$ of matrices but also its subchains. For each i, j, $1 \le i \le j \le n$, we denote by $M(i, j)$ the subchain $A_i \times A_{i+1} \times \cdots \times A_j$. We describe inductively the set $P(i, j)$ of parenthesizations for $M(i, j)$. We denote $P(i, i) = \{A_i\}$ for $i = 1, \ldots, n$. For $i < j$, we denote $P(i, j) = \bigcup_{k=i}^{j-1} P(i, k, j)$, where $P(i, k, j) = \{(p_1 \times p_2) : p_1 \in P(i, k), p_2 \in P(k + 1, j)\}$. Set $U = \bigcup_{1 \le i \le j \le n} P(i, j)$.

We now define a circuit S without repetitions over the set U, which describes the set $P(1, n)$ of parenthesizations for $M(1, n)$. For each i, $1 \le i \le n$, this circuit contains the input node $v(i, i)$ labeled with the set $\{A_i\}$. For each i, j, $1 \le i < j \le n$, the circuit S contains the unifying node $v(i, j)$ labeled with the symbol $\cup$. For each i, k, j, $1 \le i \le k < j \le n$, the circuit S contains the functional node $v(i, k, j)$ labeled with the function $\varrho(x_1, x_2)$. For any two parenthesizations p_1 and p_2, $\varrho(p_1, p_2) = (p_1 \times p_2)$. The unifying node $v(i, j)$ has $j - i$ entering edges leaving the nodes $v(i, k, j)$, $i \le k \le j - 1$. The functional node $v(i, k, j)$ has two entering edges leaving the nodes $v(i, k)$ and $v(k + 1, j)$. These edges are labeled with numbers 1 and 2, respectively (we omit these numbers in the example). The node $v(1, n)$ is the output of S and is labeled with $*$. One can show that $S(v(i, k, j)) = P(i, k, j)$ for any $i, k, j, 1 \le i \le k < j \le n$, and $S(v(i, j)) = P(i, j)$ for any $i, j, 1 \le i \le j \le n$.

The described circuit will be considered as the preliminary version of the circuit S. The final version of S is obtained from the preliminary version of S by the removal of all nodes such that there is no directed path from the considered node to the output. We also remove all edges that are incident to the removed nodes. Later, S without mentioning the version denotes the final version of the circuit S.

The preliminary version of the circuit S contains n input nodes, at most n^3 functional nodes with two entering edges each, and at most n^2 unifying nodes with at most n entering edges each, i.e., this circuit contains at most $n + n^2 + n^3$ nodes and at most $3n^3$ edges.

We consider matrices A_i and A_j with different indices i and j as different matrices. It is easy to check that in this case $S(v_1) \cap S(v_2) = \emptyset$ for any nodes $v_1, v_2 \in V_i(S) \cup V_f(S)$ such that $v_1 \neq v_2$, and the function $\varrho(x_1, x_2)$ is injective. Therefore S is a circuit without repetitions.

We implement a parenthesization from $P(1, n)$ using either one processor π or n processors $\pi_1, \pi_2, \ldots, \pi_n$. In the latter case, initially, the processor π_i contains the matrix A_i, $i = 1, \ldots, n$. Suppose we should compute $(p_1 \times p_2)$, where p_1 is computed by the processor π_{t_1} and p_2 is computed by the processor π_{t_2}. At this point, there are two possibilities, i.e., either the processor π_{t_1} sends the matrix corresponding to p_1 to the processor π_{t_2} or vice versa, where the receiving processor computes the product of matrices corresponding to p_1 and p_2.

For $t = 1, 2, 3$, we now describe a cost function $\Psi^{(t)}$, which associates to each parenthesization $p \in U$ a nonnegative integer $\Psi^{(t)}(p)$. The function $\Psi^{(t)}$ is given by the function $\psi^{(t)}$ and the functions $\psi^{(t)}_{v(i,k,j)}$, $1 \le i \le k < j \le n$. For

$t = 1, 2, 3$, $\psi^{(t)}(x) = 0$ since when we have only one matrix A_i we not need to do any computations. We will explain the considered definitions using a parenthesization $p = (p_1 \times p_2)$, where $p \in P(i, k, j)$, $p_1 \in P(i, k)$, and $p_2 \in P(k + 1, j)$, $1 \le i \le k < j \le n$. Denote

$$B = \{x + y, x \cdot y, \min(x, y), \max(x, y)\} .$$

1. Cost function $\Psi^{(1)}$, where $\psi^{(1)}(x)=0$ and $\psi^{(1)}_{v(i,k,j)}(x_1, x_2)=x_1 + x_2 + m_{i-1}m_k m_j$ for any $i, k, j, 1 \le i \le k < j \le n$. Parenthesizations p_1 and p_2 represent matrices of dimensions $m_{i-1} \times m_k$ and $m_k \times m_j$, respectively. For multiplication of these matrices we need to make $m_{i-1}m_k m_j$ scalar multiplications. So $\Psi^{(1)}(p)$ is the total number of scalar multiplications required by one or n processors to compute the product $A_i \times \cdots \times A_j$ according to the parenthesization $p = (p_1 \times p_2)$. We should add that $\Psi^{(1)}(p)$ can be considered as the time complexity of computation of p (when we count only scalar multiplications) using one processor. To compute $\psi^{(1)}$, we do not need any operations, while to compute $\psi^{(1)}_{v(i,k,j)}$, we need four operations from B. It is clear that $\Psi^{(1)}$ is a strictly increasing nonnegative integer cost function.
2. Cost function $\Psi^{(2)}$, where $\psi^{(2)}(x) = 0$ and $\psi^{(2)}_{v(i,k,j)}(x_1, x_2) = \max\{x_1, x_2\} + m_{i-1}m_k m_j$ for any i, k, j, $1 \le i \le k < j \le n$. This cost function describes the time complexity of computation of p (when we count only scalar multiplications) using n processors. To compute $\psi^{(2)}$, we do not need any operations, while to compute $\psi^{(2)}_{v(i,k,j)}$, we need four operations from B. The cost function $\Psi^{(2)}$ is an increasing nonnegative integer cost function.
3. Cost function $\Psi^{(3)}$, where $\psi^{(3)}(x) = 0$ and

$$\psi^{(3)}_{v(i,k,j)}(x_1, x_2) = x_1 + x_2 + \min\{m_{i-1}m_k, m_k m_j\}$$

for any i, k, j, $1 \le i \le k < j \le n$. This cost function describes the total cost of sending matrices between the processors when we compute p by n processors. We have the following situation: either processor π_{t_1} can send the $m_{i-1} \times m_k$ matrix to π_{t_2}, or π_{t_2} can send the $m_k \times m_j$ matrix to π_{t_1}. The number of elements in the first matrix is equal to $m_{i-1}m_k$ and the number of elements in the second matrix is equal to $m_k m_j$. To minimize the number of elements that should be sent, we must choose the minimum between $m_{i-1}m_k$ and $m_k m_j$. To compute $\psi^{(3)}$, we do not need any operations, while to compute $\psi^{(3)}_{v(i,k,j)}$, we need five operations from B. It is clear that $\Psi^{(3)}$ is a strictly increasing nonnegative integer cost function.

All these cost functions are adapted (see Sect. 2.4). They are described by formulas over the set of elementary operations B.

5.2 Time Complexity of Algorithms

The algorithms $\mathscr{A}_1$, $\mathscr{A}_2$, and $\mathscr{A}_4$ work with the scheme of the circuit S (see Sect. 2.4). The preliminary version of the circuit S is completely described. Therefore, the time complexity of the scheme of the preliminary version of S construction is linear depending on the number of its nodes and edges, i.e., $O(n^3)$. Using breadth-first search from the output node in the scheme of the preliminary version of S with the reverse direction of the edges, we can find in the scheme of the preliminary version of S all nodes such that there is no directed path from the considered node to the output and construct the scheme of S in $O(n^3)$ time. In total, we need $O(n^3)$ time to construct the scheme of S.

Note that the algorithms $\mathscr{A}_1$ and $\mathscr{A}_2$ can also work with the schemes of edge-preserving subcircuits of S obtained as a result of the algorithm $\mathscr{A}_1$ work—see Remark 3.1.

Denote $B^+ = B \cup \{x : y\}$, where $x : y$ is the operation of comparison with values $x < y$, $x = y$, and $x > y$.

Let $\mu = \max\{m_0, m_1, \ldots, m_n\}$. We now evaluate the parameter

$$N_{\Psi^{(t)}}(S) = \max\{|\Psi^{(t)}(p)| : p \in U(S)\}$$

for $t = 1, 2, 3$. One can show that $N_{\Psi^{(t)}}(S) \leq (n-1)\mu^3$ if $t \in \{1, 2\}$ and $N_{\Psi^{(t)}}(S) \leq (n-1)\mu^2$ if $t = 3$. It is clear that $N_{\Psi^{(t)}}(S) \leq n\mu^3$ for $t = 1, 2, 3$. Note that $\log_2 \mu$ is at most the length of the problem representation.

We can use the algorithm $\mathscr{A}_1$ for multi-stage optimization of parenthesizations. Using Theorem 3.1, we evaluate the number of operations from the set B^+ that the algorithm $\mathscr{A}_1$ makes when works with the scheme of an edge-preserving subcircuit T of the circuit S and a cost function $\Psi^{(t)}$, $t \in \{1, 2, 3\}$. The scheme of T contains at most n^2 unifying nodes. Each such node has at most n entering edges. So, to treat all unifying nodes, the algorithm $\mathscr{A}_1$ requires at most $2n^3$ comparisons. The scheme of T contains at most n^3 functional nodes. To treat a functional node $v(i, k, j)$ (to compute the value of $\psi^{(t)}_{v(i,k,j)}(x_1, x_2)$) the algorithm $\mathscr{A}_1$ requires at most five operations. So, to treat all functional nodes, the algorithm $\mathscr{A}_1$ requires at most $5n^3$ operations. To treat all input nodes, the algorithm $\mathscr{A}_1$ does not require any operations. Therefore, the algorithm $\mathscr{A}_1$ requires at most $7n^3$ operations from the set B^+, i.e., polynomial number of operations depending on the problem dimension n.

In the software model of computation, for the considered scheme of the circuit and cost function, the algorithm $\mathscr{A}_1$ has polynomial time complexity $O(n^3)$ depending on the problem dimension n.

Using Remark 3.2, we obtain that, for the algorithm $\mathscr{A}_1$, the number $n\mu^3$ is an upper bound on the absolute values of the numbers at the inputs of the operations $\psi^{(t)}_v$ related to the functional nodes $v \in V_f(S)$ and at the inputs of comparisons related to the unifying nodes $v \in V_u(S)$. It is easy to check that, for the algorithm $\mathscr{A}_1$, the number $n\mu^3$ is an upper bound on the absolute values of the numbers at the inputs of operations from B in formulas for the functions $\psi^{(t)}_v$. Therefore, in the integer model of computation, for the considered scheme of the circuit and the cost function $\Psi^{(t)}$, the algorithm $\mathscr{A}_1$ has the time complexity $O(n^3 \log(n\mu) \log\log(n\mu) \log\log\log(n\mu))$, which is polynomial depending on the length of the problem representation (n and $\log_2 \mu$ are at most the length of the problem representation).

We can use the algorithm $\mathscr{A}_2$ for counting the number of parenthesizations represented by the circuit S and its edge-preserving subcircuits. Using Theorem 3.2, we evaluate the number of operations from B^+ that the algorithm $\mathscr{A}_2$ makes when works with the scheme of an edge-preserving subcircuit T of the circuit S. To treat all input nodes, the algorithm $\mathscr{A}_2$ does not require any operations. The scheme of T contains at most n^2 unifying nodes. Each such node has at most n entering edges. So, to treat all unifying nodes, the algorithm $\mathscr{A}_2$ requires at most n^3 additions. The scheme of T contains at most n^3 functional nodes. Each such node has two entering edges. So, to treat all functional nodes, the algorithm $\mathscr{A}_2$ requires at most n^3 multiplications. Therefore, the algorithm $\mathscr{A}_2$ requires at most $2n^3$ operations from the set B^+, i.e., polynomial number of operations depending on the problem dimension n.

In the software model of computation, for the considered scheme of the circuit, the algorithm $\mathscr{A}_2$ has polynomial time complexity $O(n^3)$ depending on the problem dimension n.

We now evaluate the number $|S(v(1, n))|$, which is equal to the number of parenthesizations for $M(1, n)$. This value is equal to the $(n-1)$th Catalan number C_{n-1} (see [7]), where $C_n = \binom{2n}{n}/(n+1) \le 2^{2n}$ (see [2]). It is clear that the circuit S is reachable. Using Remark 3.4, we obtain that in the integer model of computation, for the considered scheme of the circuit, the algorithm $\mathscr{A}_2$ has the time complexity $O(n^4 \cdot \log n \cdot \log\log n)$, which is polynomial depending on the length of the problem representation.

We can use the algorithm $\mathscr{A}_4$ to find the set of Pareto optimal points for the problem of bi-criteria optimization of parenthesizations relative to two cost functions $\Psi^{(t)}, \Psi^{(q)}, t, q \in \{1, 2, 3\}, t \neq q$. By Proposition 4.4, $M_{\psi^{(t)},\psi^{(q)}}(S) \le n\mu^3$ if $t \in \{1, 2\}$ and $M_{\psi^{(t)},\psi^{(q)}}(S) \le n\mu^2$ if $t = 3$, where $M_{\psi^{(t)},\psi^{(q)}}(S) = \max\{|Par(t_{\psi^{(t)},\psi^{(q)}}(S, v))| : v \in V(S)\}$. Denote $M = M_{\psi^{(t)},\psi^{(q)}}(S)$.

Using Theorem 4.1, we evaluate the number of operations from B^+ that the algorithm $\mathscr{A}_4$ makes. To treat all input nodes, the algorithm $\mathscr{A}_4$ does not require any operations since $\psi^{(t)}(x) = 0$ and $\psi^{(q)}(x) = 0$. The scheme of S contains at most n^2 unifying nodes. Each such node has at most n entering edges. So, to treat all unifying nodes, the algorithm $\mathscr{A}_4$ requires at most $4n^3 M \log_2(nM)$ comparisons. The scheme of S contains at most n^3 functional nodes. Each such node has two entering edges. To compute the values of $\psi^{(t)}_{v(i,k,j)}(x_1, x_2)$ and $\psi^{(q)}_{v(i,k,j)}(x_1, x_2)$ for any functional node $v(i, k, j)$ of S we need at most 10 operations. So, to treat all functional nodes, the algorithm $\mathscr{A}_4$ requires at most $n^3(10M^2 + 8M^2 \log_2 M)$ operations. Therefore the algorithm $\mathscr{A}_4$ requires at most $4n^3 M \log_2(nM) + n^3(10M^2 + 8M^2 \log_2 M) \le 22n^3 M^2 \log_2(nM)$ operations. If $t \in \{1, 2\}$, then $M \le n\mu^3$ and $22n^3 M^2 \log_2(nM) \le 22n^5 \mu^6 \log_2(n^2 \mu^3) \le 66n^5 \mu^6 \log_2(n\mu)$. If $t = 3$, then $M \le n\mu^2$ and

$$22n^3 M^2 \log_2(nM) \le 22n^5 \mu^4 \log_2(n^2 \mu^2) \le 44n^5 \mu^4 \log_2(n\mu).$$

As a result, we obtain that the algorithm $\mathscr{A}_4$ requires at most $66n^5\mu^6\log_2(n\mu)$ operations from the set B^+ if $t \in \{1, 2\}$ and at most $44n^5\mu^4\log_2(n\mu)$ operations from the set B^+ if $t = 3$, i.e., pseudo-polynomial number of operations (polynomial depending on the dimension n and the numeric parameter μ).

In the software model of computation, for the considered scheme of the circuit and pair of cost functions, the algorithm $\mathscr{A}_4$ has the time complexity

$$O(n^5\mu^6\log(n\mu))$$

if $t \in \{1, 2\}$ and $O(n^5\mu^4\log(n\mu))$ if $t = 3$, which is pseudo-polynomial (polynomial depending on the dimension n and the numeric parameter μ).

We know that $\max(N_{\psi^{(t)}}(S), N_{\psi^{(q)}}(S)) \leq n\mu^3$. Using Remark 4.4 we obtain that, for the algorithm $\mathscr{A}_4$, the number $n\mu^3$ is an upper bound on the absolute values of the numbers at the inputs of the operations $\psi_v^{(t)}$, $\psi_v^{(q)}$ related to the functional nodes and at the inputs of comparisons related to the functional and unifying nodes of S. It is easy to check that, for the algorithm $\mathscr{A}_4$, the number $n\mu^3$ is an upper bound on the absolute values of the numbers at the inputs of operations from B in formulas for the functions $\psi_v^{(t)}$ and $\psi_v^{(q)}$. In the integer model of computation, for the considered scheme of the circuit and pair of cost functions, the algorithm $\mathscr{A}_4$ has the time complexity $O(n^5\mu^6\log(n\mu)^2\log\log(n\mu)\log\log\log(n\mu))$ if $t \in \{1, 2\}$ and

$$O(n^5\mu^4\log(n\mu)^2\log\log(n\mu)\log\log\log(n\mu))$$

if $t = 3$, which is pseudo-polynomial (polynomial depending on the length of the problem representation and the numeric parameter μ).

5.3 Example

We consider an example of matrix chain multiplication. Let A_1, A_2, A_3, and A_4 be four matrices with dimensions 4×5, 5×7, 7×5, and 5×4, respectively, i.e., $m_0 = 4$, $m_1 = 5$, $m_2 = 7$, $m_3 = 5$, and $m_4 = 4$.

The circuit S_{MCM}, which describes all parenthesizations for this chain of matrices, is presented in Fig. 5.1, where rectangles represent functional nodes and ovals represent input and unifying nodes. There are four input nodes $v(1, 1)$, $v(2, 2)$, $v(3, 3)$, and $v(4, 4)$, which are labeled with sets $\{A_1\}$, $\{A_2\}$, $\{A_3\}$, and $\{A_4\}$, respectively. Each unifying node is labeled with the symbol $\cup$, and each functional node is labeled with the function $\varrho(x_1, x_2)$. The node $v(1, 4)$ is the output of the circuit S_{MCM}.

We will encounter circuits with similar structure when considering binary search trees and convex polygon triangulations. These three types of objects are closely related to each other [4].

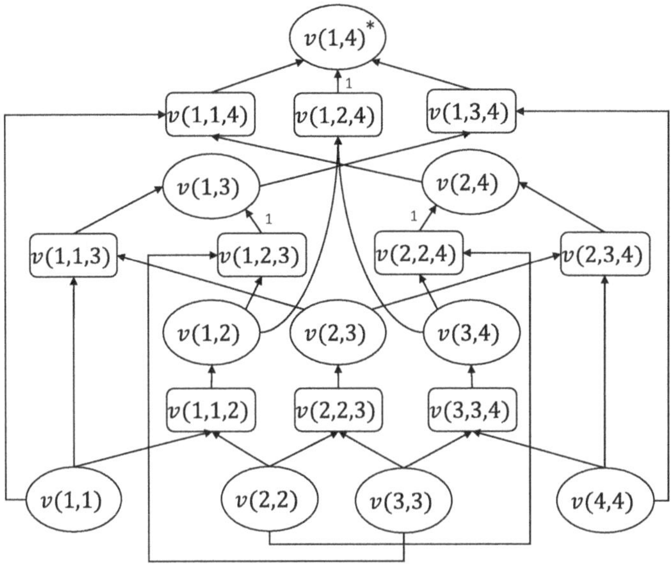

Fig. 5.1 Circuit S_{MCM} for matrix chain multiplication problem

We consider multi-stage optimization of parenthesizations relative to the cost functions $\Psi^{(1)}$ and $\Psi^{(3)}$. If we remove from the circuit S_{MCM} depicted in Fig. 5.1 all edges labeled with the number 1, then we obtain the circuits $S_{MCM}^{\Psi^{(1)}}$ and $S_{MCM}^{\Psi^{(1)},\Psi^{(3)}}$, which are equal. These circuits describe two parenthesizations:

$$(A_1 \times ((A_2 \times A_3) \times A_4)) ,$$
$$((A_1 \times (A_2 \times A_3)) \times A_4) .$$

Results of bi-criteria optimization of parenthesizations relative to the cost functions $\Psi^{(1)}$ and $\Psi^{(3)}$ can be found in Fig. 5.2, where we attached the corresponding set of Pareto optimal points to each node of the circuit S_{MCM}.

5.4 Experiments

We experimentally tested various optimization scenarios for parenthesizations for a chain of matrices $A_1 \times A_2 \times \cdots \times A_n$. We randomly generated dimensions $m_0, \ldots, m_n$ of these matrices from different intervals, for example, $[1, 10]$ or $[1, 100]$, with uniform distribution.

Table 5.1 shows, for $n = 10, 20, \ldots, 100$, the average number of parenthesizations described by the circuit before optimization, after optimization relative to $\Psi^{(1)}$, relative to $\Psi^{(2)}$, and relative to $\Psi^{(3)}$ with matrices dimensions from the intervals $[1, 10]$ or $[1, 100]$. With the increase of the matrices dimensions ($[1, 10]$ versus

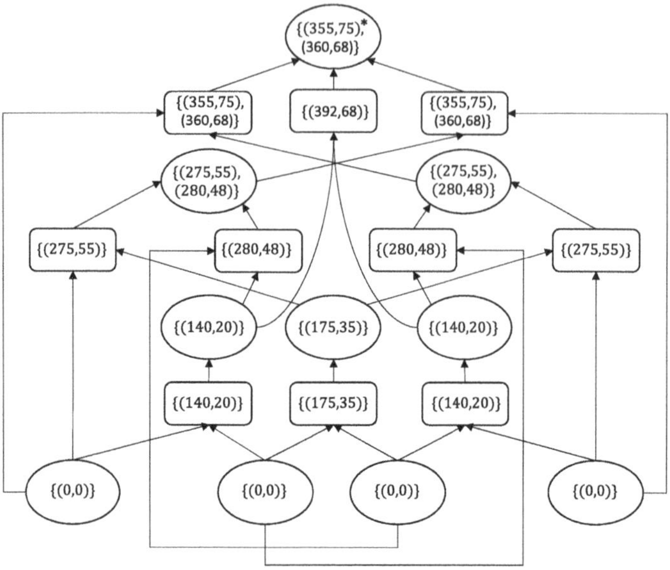

Fig. 5.2 Circuit S_{MCM} for matrix chain multiplication problem with the sets of Pareto optimal points attached to nodes

Table 5.1 Average number of parenthesizations described by the circuit before optimization, after optimization relative to $\Psi^{(1)}$, relative to $\Psi^{(2)}$, and relative to $\Psi^{(3)}$ with matrices dimensions from the intervals [1, 10] or [1, 100] among 100 trials

n	$\emptyset$	[1, 10]			[1, 100]		
		$\Psi^{(1)}$	$\Psi^{(2)}$	$\Psi^{(3)}$	$\Psi^{(1)}$	$\Psi^{(2)}$	$\Psi^{(3)}$
10	4.86×10^{03}	1.27	1.21	17.61	1.00	1.00	1.62
20	1.77×10^{09}	3.51	1.98	1.89×10^{04}	1.00	1.00	7.21
30	1.00×10^{15}	10.79	2.80	3.06×10^{06}	1.02	1.08	54.73
40	6.80×10^{20}	64.16	5.32	2.29×10^{09}	1.05	1.09	463.61
50	5.10×10^{26}	336.74	9.59	3.45×10^{11}	1.07	1.16	1.38×10^{03}
60	4.06×10^{32}	2.33×10^{03}	19.72	1.38×10^{14}	1.15	1.20	3.42×10^{03}
70	3.37×10^{38}	1.26×10^{04}	28.65	6.40×10^{16}	1.18	1.20	1.14×10^{04}
80	2.89×10^{44}	1.69×10^{05}	42.90	3.10×10^{19}	1.17	1.27	1.78×10^{04}
90	2.54×10^{50}	7.32×10^{05}	62.77	2.08×10^{22}	1.13	1.29	4.59×10^{04}
100	2.28×10^{56}	7.82×10^{07}	138.11	1.13×10^{26}	1.15	1.48	8.07×10^{04}

[1, 100]), the number of optimal parenthesizations decreased on average regardless of the cost function. For all scenarios, there were more optimal parenthesizations relative to $\Psi^{(3)}$ on average than optimal relative to $\Psi^{(1)}$ or $\Psi^{(2)}$.

Table 5.2 shows, for $n = 10, 20, \ldots, 100$, the average number of parenthesizations described by the circuit after optimization relative to $\Psi^{(1)}, \Psi^{(2)}$, relative to $\Psi^{(2)}, \Psi^{(1)}$, relative to $\Psi^{(1)}, \Psi^{(3)}$, relative to $\Psi^{(3)}, \Psi^{(1)}$, relative to $\Psi^{(2)}, \Psi^{(3)}$, and

Table 5.2 Average number of parenthesizations described by the circuit after optimization relative to $\Psi^{(1)}, \Psi^{(2)}$, relative to $\Psi^{(2)}, \Psi^{(1)}$, relative to $\Psi^{(1)}, \Psi^{(3)}$, relative to $\Psi^{(3)}, \Psi^{(1)}$, relative to $\Psi^{(2)}, \Psi^{(3)}$, and relative to $\Psi^{(3)}, \Psi^{(2)}$ with matrices dimensions from the interval [1, 10] among 100 trials

n	$\Psi^{(1)}, \Psi^{(2)}$	$\Psi^{(2)}, \Psi^{(1)}$	$\Psi^{(1)}, \Psi^{(3)}$	$\Psi^{(3)}, \Psi^{(1)}$	$\Psi^{(2)}, \Psi^{(3)}$	$\Psi^{(3)}, \Psi^{(2)}$
10	1.08	1.08	1.25	1.25	1.15	1.05
20	1.29	1.16	3.51	3.51	1.61	1.37
30	1.68	1.24	10.73	10.73	2.41	1.53
40	2.13	1.42	64.16	64.16	3.86	2.24
50	2.78	1.52	336.74	336.74	7.17	4.04
60	4.33	1.79	2.33×10^{03}	2.33×10^{03}	14.51	5.07
70	5.33	2.01	1.26×10^{04}	1.26×10^{04}	20.83	6.50
80	7.43	2.23	1.69×10^{05}	1.69×10^{05}	26.52	11.80
90	9.76	2.63	7.32×10^{05}	7.32×10^{05}	38.75	12.80
100	13.7	2.82	7.82×10^{07}	7.82×10^{07}	86.04	23.89

Table 5.3 Average value of $\Psi^{(1)}$ after optimization relative to $\Psi^{(1)}$ and relative to $\Psi^{(3)}, \Psi^{(1)}$, and average value of $\Psi^{(3)}$ after optimization relative to $\Psi^{(3)}$, relative to $\Psi^{(1)}, \Psi^{(3)}$, and relative to $\Psi^{(2)}, \Psi^{(3)}$ with matrices dimensions from the interval [1, 10] among 100 trials

n	$\Psi^{(1)}$	$\Psi^{(3)}, \Psi^{(1)}$	$\Psi^{(3)}$	$\Psi^{(1)}, \Psi^{(3)}$	$\Psi^{(2)}, \Psi^{(3)}$
10	443.41	443.57	73.72	73.80	85.57
20	668.50	668.69	117.80	117.95	150.90
30	930.68	930.77	167.75	167.90	224.17
40	1.20×10^{03}	1.20×10^{03}	218.33	218.33	290.94
50	1.49×10^{03}	1.49×10^{03}	273.82	273.82	362.33
60	1.80×10^{03}	1.80×10^{03}	329.77	329.77	444.31
70	2.07×10^{03}	2.07×10^{03}	381.19	381.19	517.52
80	2.38×10^{03}	2.38×10^{03}	437.60	437.60	588.57
90	2.67×10^{03}	2.67×10^{03}	492.51	492.51	657.09
100	2.96×10^{03}	2.96×10^{03}	546.59	546.59	727.03

relative to $\Psi^{(3)}, \Psi^{(2)}$ with matrices dimensions from the interval [1, 10]. The vast majority of parenthesizations optimal relative to $\Psi^{(1)}$ also were optimal relative to $\Psi^{(3)}$. More parenthesizations were optimal relative to $\Psi^{(2)}, \Psi^{(3)}$ on average than to $\Psi^{(3)}, \Psi^{(2)}$.

Table 5.3 shows the average value of $\Psi^{(1)}$ after optimization relative to $\Psi^{(1)}$ and relative to $\Psi^{(3)}, \Psi^{(1)}$, and the average value of $\Psi^{(3)}$ after optimization relative to $\Psi^{(3)}$, relative to $\Psi^{(1)}, \Psi^{(3)}$, and relative to $\Psi^{(2)}, \Psi^{(3)}$. The values in the columns $\Psi^{(1)}$ and $\Psi^{(3)}, \Psi^{(1)}$ are very similar. The same situation is with the columns $\Psi^{(3)}$ and $\Psi^{(1)}, \Psi^{(3)}$. The values in the columns $\Psi^{(3)}$ and $\Psi^{(2)}, \Psi^{(3)}$ are noticeably different. It means that the initial optimization relative to $\Psi^{(2)}$ significantly reduces the possibilities for the optimization relative to $\Psi^{(3)}$.

Table 5.4 Number of Pareto optimal points (in format ${}_{\min}average_{\max}$ among 100 trials) for bi-criteria optimization problem relative to $\Psi^{(1)}, \Psi^{(2)}$. Matrices dimensions are from the interval $[1, k]$

n	k				
	5	10	15	20	25
10	${}_{1}3.01_{8}$	${}_{1}3.30_{12}$	${}_{1}3.44_{12}$	${}_{1}3.01_{14}$	${}_{1}2.91_{11}$
15	${}_{1}5.28_{26}$	${}_{1}6.98_{25}$	${}_{1}5.95_{28}$	${}_{1}6.18_{19}$	${}_{1}6.45_{22}$
20	${}_{1}5.18_{17}$	${}_{1}12.89_{72}$	${}_{1}12.02_{47}$	${}_{1}13.78_{70}$	${}_{1}10.88_{51}$
25	${}_{2}9.00_{45}$	${}_{1}13.99_{65}$	${}_{1}20.72_{124}$	${}_{1}19.06_{96}$	${}_{1}18.17_{65}$
30	${}_{1}10.76_{92}$	${}_{2}21.56_{94}$	${}_{3}31.41_{176}$	${}_{1}28.85_{178}$	${}_{1}31.34_{205}$
35	${}_{1}13.32_{84}$	${}_{2}28.22_{146}$	${}_{1}41.20_{212}$	${}_{1}41.37_{200}$	${}_{1}44.11_{293}$
40	${}_{1}13.68_{77}$	${}_{1}37.97_{191}$	${}_{2}56.96_{273}$	${}_{1}67.39_{333}$	${}_{2}60.45_{375}$
45	${}_{1}14.69_{53}$	${}_{1}39.43_{207}$	${}_{2}58.25_{290}$	${}_{2}77.10_{365}$	${}_{5}85.32_{349}$
50	${}_{2}15.12_{69}$	${}_{1}41.93_{321}$	${}_{2}87.90_{428}$	${}_{3}87.48_{389}$	${}_{1}108.11_{655}$

Table 5.5 Number of Pareto optimal points (in format ${}_{\min}average_{\max}$ among 100 trials) for bi-criteria optimization problem relative to $\Psi^{(1)}, \Psi^{(3)}$. Matrices dimensions are from the interval $[1, k]$

n	k				
	5	10	15	20	25
10	${}_{1}1.00_{1}$	${}_{1}1.05_{4}$	${}_{1}1.06_{2}$	${}_{1}1.09_{5}$	${}_{1}1.08_{6}$
15	${}_{1}1.00_{1}$	${}_{1}1.06_{2}$	${}_{1}1.01_{2}$	${}_{1}1.01_{2}$	${}_{1}1.11_{4}$
20	${}_{1}1.00_{1}$	${}_{1}1.00_{1}$	${}_{1}1.02_{2}$	${}_{1}1.11_{4}$	${}_{1}1.15_{6}$
25	${}_{1}1.00_{1}$	${}_{1}1.00_{1}$	${}_{1}1.03_{2}$	${}_{1}1.07_{3}$	${}_{1}1.05_{4}$
30	${}_{1}1.00_{1}$	${}_{1}1.00_{1}$	${}_{1}1.04_{4}$	${}_{1}1.12_{4}$	${}_{1}1.04_{2}$
35	${}_{1}1.00_{1}$	${}_{1}1.01_{2}$	${}_{1}1.05_{4}$	${}_{1}1.03_{3}$	${}_{1}1.05_{4}$
40	${}_{1}1.00_{1}$	${}_{1}1.00_{1}$	${}_{1}1.00_{1}$	${}_{1}1.00_{1}$	${}_{1}1.07_{4}$
45	${}_{1}1.00_{1}$	${}_{1}1.00_{1}$	${}_{1}1.01_{2}$	${}_{1}1.01_{2}$	${}_{1}1.08_{4}$
50	${}_{1}1.00_{1}$	${}_{1}1.00_{1}$	${}_{1}1.00_{1}$	${}_{1}1.00_{1}$	${}_{1}1.00_{1}$

We also consider bi-criteria optimization of parenthesizations. Tables 5.4 and 5.5 contain information about the number of Pareto optimal points for various bi-criteria optimization problems. Table 5.4 shows the minimum, maximum, and the average number of Pareto optimal points for bi-criteria optimization problem relative to $\Psi^{(1)}, \Psi^{(2)}$ for $n = 10, 15, \ldots, 50$ with matrices dimensions from the interval $[1, k]$, $k = 5, 10, \ldots, 25$. Among 100 trials, there were at most 655 and at least one Pareto optimal points relative to $\Psi^{(1)}, \Psi^{(2)}$, while on average, there were at least 2.91 and at most 108.11 Pareto optimal points for all scenarios.

Table 5.6 shows the number of cases with totally optimal parenthesizations relative to $\Psi^{(1)}$ and $\Psi^{(2)}$ among 100 trials with matrices dimensions from the interval $[1, k]$, $k = 5, 10, \ldots, 25$. For $n = 10$, there were at least 15 cases with totally optimal

Table 5.6 Number of cases with totally optimal parenthesizations relative to $\Psi^{(1)}$ and $\Psi^{(2)}$ among 100 trials. Matrices dimensions are from the interval $[1, k]$

n	k				
	5	10	15	20	25
10	15	21	20	29	27
15	9	6	14	9	13
20	5	4	7	6	9
25	0	1	4	2	4
30	2	0	0	1	2
35	1	0	2	1	1
40	1	3	0	1	0
45	2	1	0	0	0
50	0	2	0	0	1

Table 5.7 Number of cases with totally optimal parenthesizations relative to $\Psi^{(1)}$ and $\Psi^{(3)}$ among 100 trials. Matrices dimensions are from the interval $[1, k]$

n	k				
	5	10	15	20	25
10	100	97	94	95	96
15	100	94	99	99	92
20	100	100	98	91	92
25	100	100	97	94	97
30	100	100	98	94	96
35	100	99	97	98	97
40	100	100	100	100	96
45	100	100	99	99	94
50	100	100	100	100	100

parenthesizations (among 100 trials). For $n = 50$, there were three scenarios ($k = 5, 15, 20$) without any totally optimal parenthesizations. The set of Pareto optimal points for a bi-criteria optimization of parenthesizations relative to $\Psi^{(1)}$ and $\Psi^{(2)}$ ($n = 70$, dimensions are from $[1, 10]$) is depicted in Fig. 5.3.

Table 5.5 shows the minimum, maximum, and the average number of Pareto optimal points for bi-criteria optimization problem relative to $\Psi^{(1)}, \Psi^{(3)}$ for $n = 10, 15, \ldots, 50$ with matrices dimensions from the interval $[1, k]$, $k = 5, 10, \ldots, 25$. On average, there were at most 1.15 Pareto optimal points, and the maximum number of Pareto optimal points was six among all tested scenarios.

For each case with $k = 5$, there were totally optimal parenthesizations relative to $\Psi^{(1)}, \Psi^{(3)}$ (see Table 5.7). The set of Pareto optimal points for a bi-criteria optimization of parenthesizations relative to $\Psi^{(1)}$ and $\Psi^{(3)}$ ($n = 70$, dimensions are from $[1, 10]$) is depicted in Fig. 5.4.

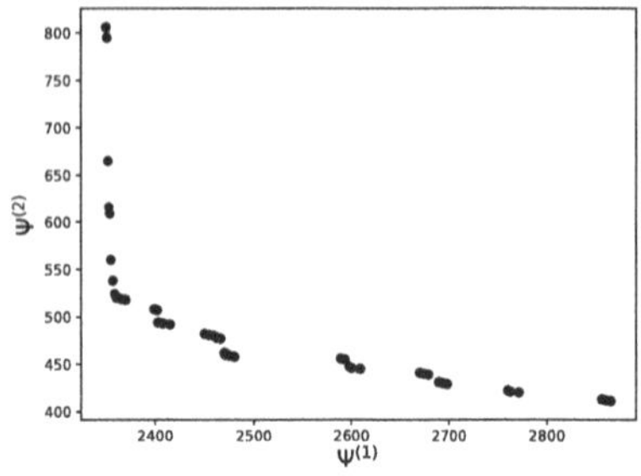

Fig. 5.3 Set of Pareto optimal points for a bi-criteria optimization of parenthesizations relative to $\Psi^{(1)}$ and $\Psi^{(2)}$. Matrices dimensions are from the interval $[1, 10]$, $n = 70$

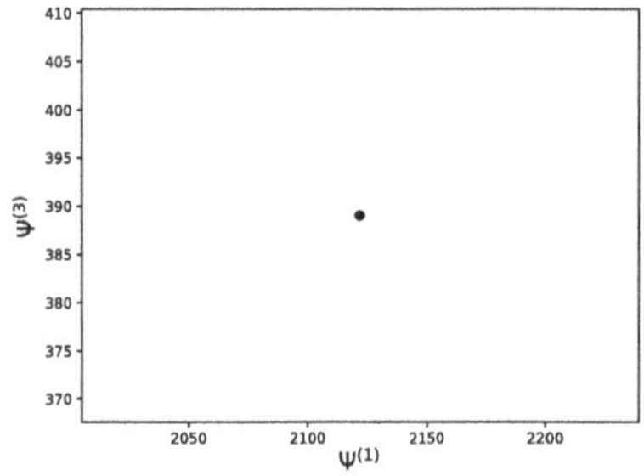

Fig. 5.4 Set of Pareto optimal points for a bi-criteria optimization of parenthesizations relative to $\Psi^{(1)}$ and $\Psi^{(3)}$. Matrices dimensions are from the interval $[1, 10]$, $n = 70$

References

1. Chikalov, I., Hussain, S., Moshkov, M.: Sequential optimization of matrix chain multiplication relative to different cost functions. In: Cerná, I., Gyimóthy, T., Hromkovic, J., Jeffery, K.G., Královic, R., Vukolic, M., Wolf, S. (eds.) 37th Conference on Current Trends in Theory and Practice of Computer Science, SOFSEM 2011, Nový Smokovec, Slovakia, January 22–28, 2011. Lecture Notes in Computer Science, vol. 6543, pp. 157–165. Springer (2011)
2. Dutton, R.D., Brigham, R.C.: Computationally efficient bounds for the Catalan numbers. Eur. J. Comb. **7**(3), 211–213 (1986)
3. Godbole, S.: On efficient computation of matrix chain products. IEEE Trans. Comput. **22**(9), 864–866 (1973)
4. Hu, T.C., Shing, M.T.: Computation of matrix chain products Part I. SIAM J. Comput. **11**(2), 362–373 (1982)
5. Hu, T.C., Shing, M.T.: Computation of matrix chain products Part II. SIAM J. Comput. **13**(2), 228–251 (1984)
6. Mankowski, M., Moshkov, M.: Dynamic programming bi-criteria combinatorial optimization. Discret. Appl. Math. **284**, 513–533 (2020)
7. Stanley, R.P.: Enumerative Combinatorics. Cambridge Studies in Advanced Mathematics, vol. 2. Cambridge University Press (1999)

Chapter 6
Global Sequence Alignment

The sequence alignment problem is a generalization of the edit distance problem (see Levenshtein [4]), which measures the difference between two sequences by the minimum number of removals, insertions, and substitutions of symbols required to obtain one sequence from the another. The solution to the sequence alignment problem helps to identify biological similarities in long sequences of DNA, RNA, and proteins.

There are two different approaches to sequence alignment: the global alignment and the local alignment. The local alignment aims to align parts of two sequences that are very long. On the other hand, global alignment techniques try to align the whole sequences. Needleman and Wunsch [5], and Smith and Waterman [6] present general dynamic programming algorithms for global and local alignment problems, respectively. In general, the alignment problems are not limited to paring only two sequences; the multiple sequence alignment is widely studied [2].

In this chapter, we study the global sequence alignment problem. We consider a conventional circuit without repetitions for the representation of all alignments of given sequences and define a class of cost functions for this circuit. We evaluate the number of operations and time required by the algorithms for multi-stage optimization of alignments, for counting the optimal alignments, and for bi-criteria optimization of alignments. Note that an algorithm for multi-stage optimization of alignments relative to a sequence of cost functions was proposed in [1]. We also consider an example and show the results of computer experiments with randomly generated sequences. In the example and experiments, we use two cost functions from the considered class: the number of indels—the number of matches of letters and gaps in an alignment and the number of mismatches—the number of matches of different letters in an alignment (we consider here the notion of gap in the same way as in [3]).

M. Mankowski and M. Moshkov, *Dynamic Programming Multi-Objective Combinatorial Optimization*, Studies in Systems, Decision and Control 331,
https://doi.org/10.1007/978-3-030-63920-4_6

6.1 Definition of Circuit and Cost Functions

Let $X = x_1x_2 \cdots x_m$, $Y = y_1y_2 \cdots y_n$ be nonempty sequences (words) over a finite alphabet Σ. The numbers m and n are the dimensions of the problems under consideration related to the alignment of these sequences. To define a cost function Ψ for the sequence alignment problem, we must specify $mn + m + n$ nonnegative integers $w_\Psi(x_i, \sqcup)$, $w_\Psi(\sqcup, y_j)$, and $w_\Psi(x_i, y_j)$, $1 \le i \le m$, $1 \le j \le n$ that describe the costs of alignment of symbols and alignment of symbols and gaps. To describe an instance of the problem, we need one or two cost functions. Binary representations of $mn + m + n$ or $2(mn + m + n)$ numbers defining one or two cost functions form the representation of the problem instance (later, we will omit the word instance). Since m and n are at most the number of problem parameters, m and n are at most the length of the problem representation.

For each $i \in \{0, 1, \ldots, m\}$ and $j \in \{0, 1, \ldots, n\}$, we define the set $P(i, j)$ of sequence alignments of prefixes $X_i = x_1 \cdots x_i$ and $Y_j = y_1 \cdots y_j$ of sequences X and Y. If $i = 0$ and $j = 0$, then we have empty prefixes denoted by λ.

Each alignment (α, β) of X_i and Y_j is a pair of sequences of equal length (the length of alignment) obtained from prefixes X_i and Y_j by insertion of gaps "$\sqcup$" such that there is no gap in the same position in both sequences. We have $P(0, 0) = \{(\lambda, \lambda)\}$, where λ is the empty sequence, $P(i, 0) = \{(x_1 \cdots x_i, \sqcup \cdots \sqcup)\}$, and $P(0, j) = \{(\sqcup \cdots \sqcup, y_1 \cdots y_j)\}$ for any $i \in \{1, \ldots, m\}$ and $j \in \{1, \ldots, n\}$. Furthermore, for any $i \in \{1, \ldots, m\}$ and $j \in \{1, \ldots, n\}$,

$$\begin{aligned} P_1(i, j) &= \{(\alpha x_i, \beta \sqcup) : (\alpha, \beta) \in P(i-1, j)\} , \\ P_2(i, j) &= \{(\alpha \sqcup, \beta y_j) : (\alpha, \beta) \in P(i, j-1)\} , \\ P_3(i, j) &= \{(\alpha x_i, \beta y_j) : (\alpha, \beta) \in P(i-1, j-1)\} . \end{aligned} \tag{6.1}$$

Then $P(i, j) = P_1(i, j) \cup P_2(i, j) \cup P_3(i, j)$. Denote $U = \bigcup_{0 \le i \le m, 0 \le j \le n} P(i, j)$.

We define a circuit S without repetitions over the set U, which describes the set $P(m, n)$ of sequence alignments of X and Y.

Let $i \in \{0, 1, \ldots, m\}$ and $j \in \{0, 1, \ldots, n\}$. We now define three types of concatenation functions. Let $i > 0$ and $(\alpha, \beta) \in P(i-1, j)$. Then $\mu^{(1)}_{i,j}$ is a function from $P(i-1, j)$ to $P(i, j)$ such that $\mu^{(1)}_{i,j}((\alpha, \beta)) = (\alpha x_i, \beta \sqcup)$. Let $j > 0$ and $(\alpha, \beta) \in P(i, j-1)$. Then $\mu^{(2)}_{i,j}$ is a function from $P(i, j-1)$ to $P(i, j)$ such that $\mu^{(2)}_{i,j}((\alpha, \beta)) = (\alpha \sqcup, \beta y_j)$. Let $i > 0$, $j > 0$ and $(\alpha, \beta) \in P(i-1, j-1)$. Then $\mu^{(3)}_{i,j}$ is a function from $P(i-1, j-1)$ to $P(i, j)$ such that $\mu^{(3)}_{i,j}((\alpha, \beta)) = (\alpha x_i, \beta y_j)$.

The circuit S contains one input node $v(0, 0)$ labeled with the set $\{(\lambda, \lambda)\}$.

For each $i \in \{0, 1, \ldots, m\}$ and $j \in \{0, 1, \ldots, n\}$, $(i, j) \ne (0, 0)$, S contains the unifying node $v(i, j)$ labeled with the symbol $\cup$.

For each $i \in \{1, 2, \ldots, m\}$, $j \in \{1, 2, \ldots, n\}$, and $t \in \{1, 2, 3\}$, the circuit S contains a functional node $v_t(i, j)$ labeled with the function $\mu^{(t)}_{i,j}$. The node $v_1(i, j)$ has one entering edge leaving $v(i-1, j)$. The node $v_2(i, j)$ has one entering edge leaving $v(i, j-1)$. The node $v_3(i, j)$ has one entering edge leaving $v(i-1, j-1)$.

For each $i \in \{1, 2, \ldots, m\}$, the circuit S contains a functional node $v_1(i, 0)$ labeled with the function $\mu^{(1)}_{i,0}$. This node has one entering edge leaving $v(i-1, 0)$. For each $j \in \{1, 2, \ldots, n\}$, the circuit S contains a functional node $v_2(0, j)$ labeled with the function $\mu^{(2)}_{0,j}$. This node has one entering edge leaving $v(0, j-1)$. The layout is illustrated in Fig. 6.1.

For each $i \in \{1, 2, \ldots, m\}$ and $j \in \{1, 2, \ldots, n\}$, the unifying node $v(i, j)$ has three entering edges leaving the functional nodes $v_t(i, j)$, $t = 1, 2, 3$. For each $i \in \{1, 2, \ldots, m\}$, the unifying node $v(i, 0)$ has one entering edge leaving the functional node $v_1(i, 0)$. For each $j \in \{1, 2, \ldots, n\}$, the unifying node $v(0, j)$ has one entering edge leaving the functional node $v_2(0, j)$. The layout is illustrated in Fig. 6.1. The node $v(m, n)$ is considered as the output of S and is labeled with $*$.

One can show that $S(v(i, j)) = P(i, j)$ for $i = 0, \ldots, m$ and $j = 0, \ldots, n$,

$$S(v_1(i, j)) = P_1(i, j), \quad S(v_2(i, j)) = P_2(i, j), \quad S(v_3(i, j)) = P_3(i, j)$$

for $i = 1, \ldots, m$ and $j = 1, \ldots, n$, $S(v(i, 0)) = P(i, 0)$ for $i = 1, \ldots, m$, and $S(v(0, j)) = P(0, j)$ for $j = 1, \ldots, n$.

The described circuit will be considered as the preliminary version of the circuit S. The final version of S is obtained from the preliminary version of S by the removal of all nodes such that there is no directed path from the considered node to the output. We also remove all edges that are incident to the removed nodes. Later, S without mentioning the version denotes the final version of the circuit S.

The preliminary version of the circuit S contains one input node, $mn + m + n$ unifying nodes (each of mn nodes has three entering edges and each of $m + n$ nodes has one entering edge), and $3mn + m + n$ functional nodes (each functional node has one entering edge, and we will not label this edge with the number 1). Hence, the preliminary version of S contains $4mn + 2m + 2n + 1$ nodes and $6mn + 2m + 2n$ edges.

We now prove that S is a circuit without repetitions. It is clear that the concatenation function $\mu^{(t)}_{i,j}$ is injective, $t = 1, 2, 3$. Let v_1 and v_2 be different nodes from $V_i(S) \cup V_f(S)$. We now show that $S(v_1) \cap S(v_2) = \emptyset$. Let v_1 be the input node and v_2 be a functional node. Then $S(v_1)$ contains the only alignment of the length zero, but $S(v_2)$ does not contain the alignment of the length zero. Let v_1 and v_2 be functional nodes, $v_1 = v_{t_1}(i_1, j_1)$, and $v_2 = v_{t_2}(i_2, j_2)$. If $(i_1, j_1) = (i_2, j_2)$ and $t_1 \neq t_2$, then the sets $S(v_1)$ and $S(v_2)$ cannot share common alignments, since $S(v_1) = P_{t_1}(i_1, j_1)$, $S(v_2) = P_{t_2}(i_1, j_1)$, and $P_{t_1}(i_1, j_1) \cap P_{t_2}(i_1, j_1) = \emptyset$ (see (6.1)). Let $(i_1, j_1) \neq (i_2, j_2)$. One can show that $S(v_1) \subseteq P(i_1, j_1)$, $S(v_2) \subseteq P(i_2, j_2)$ and, evidently, $P(i_1, j_1) \cap P(i_2, j_2) = \emptyset$.

We now define the notion of a cost function Ψ for the circuit S given by a function ψ and function ψ_v for each $v \in V_f(S)$. Let $\psi((\lambda, \lambda)) = 0$. The function ψ_v, $v \in V_f(S)$, is defined by nonnegative integers $w_\Psi(x_i, \sqcup)$, $w_\Psi(\sqcup, y_j)$, and $w_\Psi(x_i, y_j)$, $1 \leq i \leq m$, $1 \leq j \leq n$. If $v = v_1(i, j)$, then $\psi_v(x) = x + w_\Psi(x_i, \sqcup)$. If $v = v_2(i, j)$, then $\psi_v(x) = x + w_\Psi(\sqcup, y_j)$. If $v = v_3(i, j)$, then $\psi_v(x) = x + w_\Psi(x_i, y_j)$. To compute ψ, we do not need any operations, while to compute ψ_v, we need only one addition. It is clear that Ψ is a strictly increasing nonnegative integer cost function. Changing

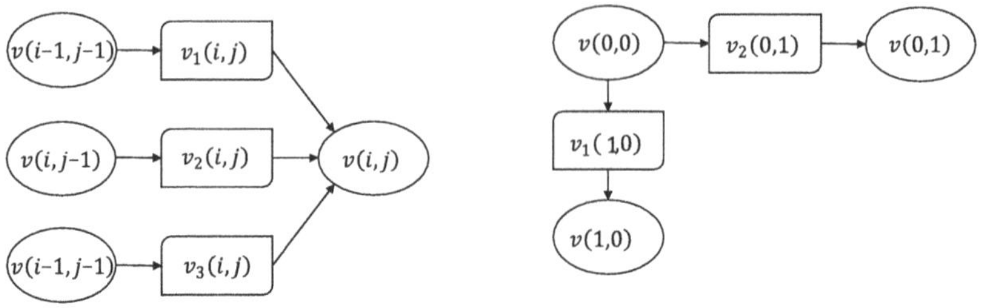

Fig. 6.1 Components of circuit S

the parameters $w_\Psi(x_i, \sqcup)$, $w_\Psi(\sqcup, y_j)$, and $w_\Psi(x_i, y_j)$, $1 \le i \le m$, $1 \le j \le n$, we get a class of cost functions for the circuit S.

Each cost function from this class is adapted (see Sect. 2.4). It is described by formulas over the set of elementary operations $B = \{x + y\}$.

For $q = 1, 2$, we describe a cost function $\Psi^{(q)}$ from the considered class, which is defined by nonnegative integers $w_{\Psi^{(q)}}(x_i, \sqcup)$, $w_{\Psi^{(q)}}(\sqcup, y_j)$, and $w_{\Psi^{(q)}}(x_i, y_j)$, $1 \le i \le m$, $1 \le j \le n$. We use functions $\Psi^{(1)}$ and $\Psi^{(2)}$ in the example and experiments.

1. Cost function $\Psi^{(1)}$ (the number of indels), where $w_{\Psi^{(1)}}(x_i, \sqcup) = 1$, $w_{\Psi^{(1)}}(\sqcup, y_j) = 1$, and $w_{\Psi^{(1)}}(x_i, y_j) = 0$, $1 \le i \le m$, $1 \le j \le n$.
2. Cost function $\Psi^{(2)}$ (the number of mismatches), where

$$w_{\Psi^{(2)}}(x_i, \sqcup) = 0,\ w_{\Psi^{(2)}}(\sqcup, y_j) = 0\ ,$$

and

$$w_{\Psi^{(2)}}(x_i, y_j) = \begin{cases} 0, & \text{if } x_i = y_j\ , \\ 1, & \text{otherwise}\ , \end{cases}$$

$1 \le i \le m$, $1 \le j \le n$.

6.2 Time Complexity of Algorithms

The algorithms $\mathscr{A}_1$, $\mathscr{A}_2$, and $\mathscr{A}_4$ work with the scheme of the circuit S (see Sect. 2.4). The preliminary version of the circuit S is completely described. Therefore, the time complexity of the scheme of the preliminary version of S construction is linear depending on the number of its nodes and edges, i.e., $O(mn)$. Using breadth-first search from the output node in the scheme of the preliminary version of S with the reverse direction of the edges, we can find in the scheme of the preliminary version of S all nodes such that there is no directed path from the considered node to the output and construct the scheme of S in $O(mn)$ time. In total, we need $O(mn)$ time to construct the scheme of S.

Note that the algorithms $\mathscr{A}_1$ and $\mathscr{A}_2$ can also work with the schemes of edge-preserving subcircuits of S obtained as a result of the algorithm $\mathscr{A}_1$ work—see Remark 3.1.

Denote $B^+ = B \cup \{x : y\}$, where $x : y$ is the operation of comparison with values $x < y$, $x = y$, and $x > y$.

Let Ψ be one of the considered cost functions specified by nonnegative integers $w_\Psi(x_i, y_j)$, $w_\Psi(x_i, \text{␣})$, and $w_\Psi(\text{␣}, y_j)$, $1 \leq i \leq m$, $1 \leq j \leq n$. Denote $\nu_\Psi = \max\left\{w_\Psi(x_i, y_j), w_\Psi(x_i, \text{␣}), w_\Psi(\text{␣}, y_j) : 1 \leq i \leq m, 1 \leq j \leq n\right\}$. One can show that

$$N_\Psi(S) \leq \nu_\Psi(m + n) .$$

It is clear that $\log_2 \nu_\Psi$ is at most the length of the problem representation if the problem uses the cost function Ψ.

We can use the algorithm $\mathscr{A}_1$ for multi-stage optimization of sequence alignments. Using Theorem 3.1, we evaluate the number of operations from B^+ that the algorithm $\mathscr{A}_1$ makes while working with the scheme of an edge-preserving subcircuit T of the circuit S and the cost function Ψ for S. The scheme of T contains one input node. We do not need any operations to process this node. The scheme of T contains at most $mn + m + n$ unifying nodes. For at most mn nodes with three entering edges, the algorithm $\mathscr{A}_1$ makes at most $5mn$ comparisons. For at most $m + n$ nodes with one entering edge, $\mathscr{A}_1$ makes at most $m + n$ comparisons. In total, at most $5mn + m + n \leq 7mn$ comparisons. The scheme contains at most $3mn + m + n$ functional nodes. For each functional node v, it is enough to make one addition to compute ψ_v. The total number of required additions is at most $3mn + m + n \leq 5mn$. The total number of operations from B^+ made by $\mathscr{A}_1$ is at most $12mn$, i.e., it is polynomial depending on the problem dimensions m and n.

In the software model of computation, for the considered scheme of the circuit and the cost function Ψ, the algorithm $\mathscr{A}_1$ has polynomial time complexity $O(mn)$ depending on the problem dimensions m and n. Using Remark 3.2, we obtain that in the integer model of computation, for the considered scheme of the circuit and cost function, the algorithm $\mathscr{A}_1$ has the time complexity $O(mn \log(\nu_\Psi(m + n)))$, which is polynomial depending on the length of the problem representation (m, n, and $\log_2 \nu_\Psi$ are at most the length of the problem representation).

To count all possible sequence alignments described by T, we apply the algorithm $\mathscr{A}_2$. We use Theorem 3.2 to evaluate the number of operations from B^+ made by $\mathscr{A}_2$. The scheme of T contains at most mn unifying nodes with three entering edges and at most $m + n$ with one entering edge. Therefore, we need at most $2mn$ additions. To treat all functional nodes and the input node, we do not need any operations. The total number of operations from B^+ made by $\mathscr{A}_2$ is at most $2mn$, i.e., it is polynomial depending on the problem dimensions m and n.

In the software model of computation, for the considered scheme of the circuit, the algorithm $\mathscr{A}_2$ has polynomial time complexity $O(mn)$ depending on the problem dimensions m and n.

We now evaluate the number $|S(v(m, n))|$, which is equal to the number of sequence alignments of X and Y. Each alignment (α, β) of X and Y is a pair of sequences of equal length (the length of alignment) obtained from X and Y by insertion of gaps "␣" such that there is no gap in the same position in both sequences. Let (α, β) be an alignment with $\alpha = \alpha_1 \cdots \alpha_k$ and $\beta = \beta_1 \cdots \beta_k$. It is clear that $k \leq m + n$. We correspond to the alignment (α, β) two tuples $\delta(\alpha) = (\delta_1, \ldots, \delta_{m+n})$ and $\sigma(\beta) = (\sigma_1, \ldots, \sigma_{m+n})$ from $\{0, 1\}^{m+n}$ in the following way. For $i = 1, \ldots, m+n$, δ_i is equal to 1 if and only if $i \leq k$ and α_i is a letter from the alphabet Σ. In the same way, σ_i is equal to 1 if and only if $i \leq k$ and β_i is a letter from the alphabet Σ. One can show that for different alignments (α', β') and (α'', β''), the pairs of tuples $(\delta(\alpha'), \sigma(\beta'))$ and $(\delta(\alpha''), \sigma(\beta''))$ are different. Therefore, the number of alignments is at most $2^{2(m+n)}$. It is clear that the circuit S is reachable. Using Remark 3.4, we obtain that in the integer model of computation, for the considered scheme of the circuit, the algorithm $\mathscr{A}_2$ has the time complexity $O(mn(m+n))$, which is polynomial depending on the length of the problem representation.

For $p = 1, 2$, let Ψ_p be a nonnegative integer cost function from the considered class given by function ψ^p and ψ^p_v, $v \in V_f(S)$. We can use the algorithm $\mathscr{A}_4$ to find the set of Pareto optimal points for the problem of bi-criteria optimization of sequence alignments of X and Y relative to Ψ_1 and Ψ_2.

By Proposition 4.4, $M_{\Psi_1,\Psi_2}(S) \leq \nu_{\Psi_1}(m+n) + 1$, where

$$M_{\Psi_1,\Psi_2}(S) = \max\{\left|Par(t_{\Psi_1,\Psi_2}(S, v))\right| : v \in V(S)\} .$$

Denote $M = M_{\Psi_1,\Psi_2}(S)$.

To evaluate the number of operations from B^+ made by the algorithm $\mathscr{A}_4$, we use Theorem 4.1. To treat the input node of the scheme of S, we do not need any operations.

The scheme of S contains at most $3mn + m + n$ functional nodes. Each node has one entering edge. It requires one addition to compute the value of the function $\psi^p_{v_t(i,j)}$, $p = 1, 2$. Therefore, the algorithm needs at most $2(3mn + m + n)M$ additions and at most $(3mn + m + n)4M \log_2 M$ comparisons to process all functional nodes. In is clear that $2(3mn + m + n)M + (3mn + m + n)4M \log_2 M \leq 30mnM \log_2(3M)$.

For each of at most mn unifying nodes with three entering edges, we make at most $12M \log_2(3M)$ comparisons. For any of at most $m + n$ unifying nodes with one entering edge, we make at most $4M \log_2 M$ comparisons. Then, to process all unifying nodes, we need at most $12mnM \log_2(3M) + 4(m+n)M \log_2 M \leq 20mnM \log_2(3M)$ comparisons.

Hence, the algorithm $\mathscr{A}_4$ makes at most $50mnM \log_2(3M)$ additions and comparisons.

Since $M \leq \nu_{\Psi_1}(m+n) + 1$, the algorithm $\mathscr{A}_4$ makes at most $50mn(\nu_{\Psi_1}(m+n) + 1)) \log_2(3\nu_{\Psi_1}(m+n) + 3))$ operations from B^+, i.e., pseudo-polynomial number of operations (polynomial depending on the dimensions m and n and the numeric parameter ν_{Ψ_1}).

In the software model of computation, for the considered scheme of the circuit and pair of cost functions, the algorithm $\mathscr{A}_4$ has the time complexity $O(mn\nu_{\Psi_1}(m+n) \log(\nu_{\Psi_1}(m+n)))$, which is pseudo-polynomial (polynomial depending on the dimensions m and n and the numeric parameter ν_{Ψ_1}). Using Remark 4.4, we obtain that in the integer model of computation, for the considered scheme of the circuit and pair of cost functions, the algorithm $\mathscr{A}_4$ has the time complexity $O(mn\nu_{\Psi_1}(m+n)\log(\max(\nu_{\Psi_1}, \nu_{\Psi_2})(m+n))^2)$, which is pseudo-polynomial (polynomial depending on the length of the problem representation and the numeric parameter ν_{Ψ_1}).

6.3 Example

We consider an example of two sequences $X = AAT$ and $Y = AT$ over the alphabet $\Sigma = \{A, T\}$.

The circuit S_{SA}, which describes all alignments of these two sequences, is presented in Fig. 6.2, where rectangles represent functional nodes and ovals represent input and unifying nodes. There is one input node $v(0, 0)$, which is labeled with the set $\{(\lambda, \lambda)\}$. Each unifying node is labeled with the symbol $\cup$, and each functional node $v_t(i, j)$ is labeled with the function $\mu_{i,j}^{(t)}$. The node $v(3, 2)$ is the output of the circuit S_{SA} and is labeled with $*$.

We consider multi-stage optimization of alignments relative to the cost functions $\Psi^{(1)}$ and $\Psi^{(2)}$. The circuit $S_{SA}^{\Psi^{(1)}}$ can be obtained from S_{SA} (depicted in Fig. 6.2) by the removal of all edges labeled with the number 1. The circuit $S_{SA}^{\Psi^{(1)},\Psi^{(2)}}$ can be obtained from S_{SA} by the removal of all edges labeled with the numbers 1 and 2.

There are 25 possible alignments represented by the circuit S_{SA}. The circuit $S_{SA}^{\Psi^{(1)}}$ represents three alignments $(AAT, A\text{\textvisiblespace}T)$, $(AAT, AT\text{\textvisiblespace})$, and $(AAT, \text{\textvisiblespace}AT)$ that are optimal relative to $\Psi^{(1)}$, while the circuit $S_{SA}^{\Psi^{(1)},\Psi^{(2)}}$ represents two alignments $(AAT, A\text{\textvisiblespace}T)$ and $(AAT, \text{\textvisiblespace}AT)$.

Results of bi-criteria optimization of alignments relative to the cost functions $\Psi^{(1)}$ and $\Psi^{(2)}$ can be found in Fig. 6.3, where we attached the corresponding set of Pareto optimal points to each node of the circuit S_{SA}.

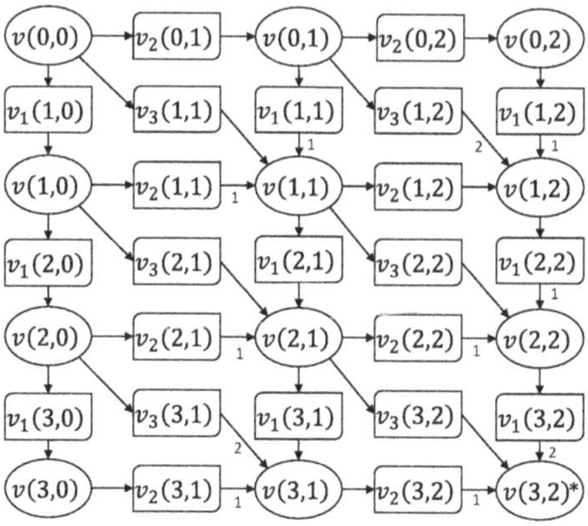

Fig. 6.2 Circuit S_{SA} for global sequence alignment

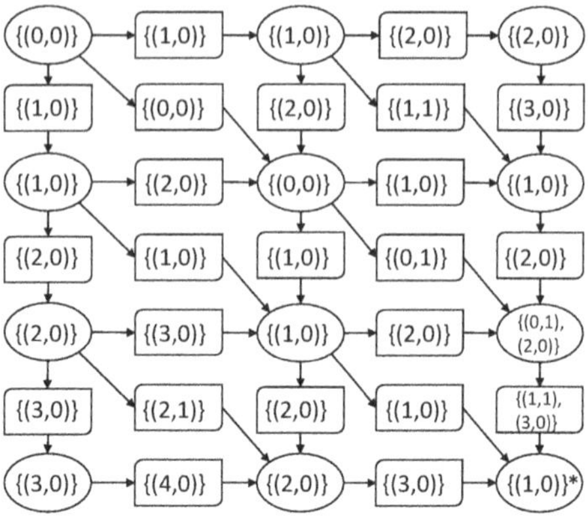

Fig. 6.3 Circuit S_{SA} for global sequence alignment with sets of Pareto optimal points attached to nodes

6.4 Experiments

We experimentally tested various optimization scenarios for global sequence alignments. We randomly generated two sequences of the lengths n and $\lceil 1.1n \rceil$ over an alphabet with the fixed number of distinct characters with uniform distribution.

Table 6.1 shows, for $n = 20, 25, \ldots, 50$, the average number of alignments described by the circuit before optimization, after optimization relative to $\Psi^{(1)}$, relative to $\Psi^{(2)}$, relative to $\Psi^{(1)}, \Psi^{(2)}$, and relative to $\Psi^{(2)}, \Psi^{(1)}$. Sequences were generated over two alphabets with 4 (A) and 23 (B) distinct characters. Regardless of the alphabet, there was identical number of alignments optimal relative to $\Psi^{(1)}$ for all tested scenarios. Comparing A and B, for the alphabet with 4 distinct characters (A), there were more alignments optimal relative to $\Psi^{(2)}$ on average, but there were less alignments optimal relative to $\Psi^{(1)}, \Psi^{(2)}$ and $\Psi^{(2)}, \Psi^{(1)}$.

Table 6.2 shows the average value of $\Psi^{(1)}$ after optimization relative to $\Psi^{(1)}$ and $\Psi^{(2)}, \Psi^{(1)}$, and the average value of $\Psi^{(2)}$ after optimization relative to $\Psi^{(2)}$ and $\Psi^{(1)}, \Psi^{(2)}$. In all tested scenarios, the value of $\Psi^{(2)}$ was 0 for sequence alignment optimization relative to $\Psi^{(2)}$. The value of $\Psi^{(2)}$ decreased in B with comparison to A for the optimization relative to $\Psi^{(1)}, \Psi^{(2)}$.

We also consider bi-criteria optimization of alignments. Table 6.3 shows the minimum, maximum, and the average number of Pareto optimal points for bi-criteria optimization problem relative to $\Psi^{(1)}, \Psi^{(2)}$ for $n = 20, 25, \ldots, 50$ with the number of distinct characters in an alphabet $k = 4, 8, 12, 16, 20, 24$. There were at least six and at most 39 Pareto optimal points in all tested scenarios. Also, there were no totally optimal alignments relative to $\Psi^{(1)}$ and $\Psi^{(2)}$ (see Table 6.4). The set of Pareto optimal points for a bi-criteria optimization of alignments relative to $\Psi^{(1)}$ and $\Psi^{(2)}$ for $n = 50$ and $k = 4$ is depicted in Fig. 6.4.

Table 6.1 Average number of sequence alignments described by the circuit before optimization, after optimization relative to $\Psi^{(1)}$, relative to $\Psi^{(2)}$, relative to $\Psi^{(1)}, \Psi^{(2)}$, and relative to $\Psi^{(2)}, \Psi^{(1)}$ among 100 trials with number of characters in the alphabet 4 (A) and 23 (B)

A

n	$\emptyset$	$\Psi^{(1)}$	$\Psi^{(2)}$	$\Psi^{(1)}, \Psi^{(2)}$	$\Psi^{(2)}, \Psi^{(1)}$
20	1.39×10^{15}	231.00	5.51×10^{12}	11.61	2.39×10^{04}
25	1.91×10^{19}	3.28×10^{03}	1.92×10^{16}	29.21	1.19×10^{05}
30	1.20×10^{23}	5.46×10^{03}	3.20×10^{19}	36.19	3.02×10^{06}
35	1.71×10^{27}	8.23×10^{04}	1.08×10^{23}	66.26	1.51×10^{07}
40	1.10×10^{31}	1.36×10^{05}	1.82×10^{26}	98.37	2.41×10^{07}
45	1.59×10^{35}	2.12×10^{06}	6.61×10^{29}	238.71	5.65×10^{08}
50	1.04×10^{39}	3.48×10^{06}	1.14×10^{33}	238.93	1.29×10^{10}

B

n	$\emptyset$	$\Psi^{(1)}$	$\Psi^{(2)}$	$\Psi^{(1)}, \Psi^{(2)}$	$\Psi^{(2)}, \Psi^{(1)}$
20	1.39×10^{15}	231.00	8.30×10^{11}	63.31	6.72×10^{06}
25	1.91×10^{19}	3.28×10^{03}	1.64×10^{15}	273.53	1.19×10^{08}
30	1.20×10^{23}	5.46×10^{03}	1.66×10^{18}	444.84	1.38×10^{10}
35	1.71×10^{27}	8.23×10^{04}	3.32×10^{21}	2.51×10^{03}	4.69×10^{10}
40	1.10×10^{31}	1.36×10^{05}	3.69×10^{24}	3.25×10^{03}	1.42×10^{12}
45	1.59×10^{35}	2.12×10^{06}	7.69×10^{27}	1.34×10^{04}	1.27×10^{14}
50	1.04×10^{39}	3.48×10^{06}	8.67×10^{30}	2.74×10^{04}	2.06×10^{15}

Table 6.2 Average value of $\Psi^{(1)}$ after optimization relative to $\Psi^{(1)}$ and $\Psi^{(2)}, \Psi^{(1)}$, and average value of $\Psi^{(2)}$ after optimization relative to $\Psi^{(2)}$ and $\Psi^{(1)}, \Psi^{(2)}$ among 100 trials with number of characters in the alphabet 4 (A) and 23 (B)

A

n	$\Psi^{(1)}$	$\Psi^{(2)}, \Psi^{(1)}$	$\Psi^{(2)}$	$\Psi^{(1)}, \Psi^{(2)}$
20	2.00	30.60	0.00	17.87
25	3.00	38.26	0.00	21.86
30	3.00	45.62	0.00	26.68
35	4.00	53.20	0.00	30.94
40	4.00	59.70	0.00	35.44
45	5.00	67.40	0.00	39.42
50	5.00	73.96	0.00	43.95

B

n	$\Psi^{(1)}$	$\Psi^{(2)}, \Psi^{(1)}$	$\Psi^{(2)}$	$\Psi^{(1)}, \Psi^{(2)}$
20	2.00	18.06	0.00	12.11
25	3.00	22.16	0.00	14.33
30	3.00	25.82	0.00	17.79
35	4.00	29.70	0.00	19.52
40	4.00	33.56	0.00	23.38
45	5.00	37.76	0.00	25.02
50	5.00	41.58	0.00	28.58

Table 6.3 Number of Pareto optimal points (in format $_{min}average_{max}$ among 100 trials) for bi-criteria optimization problem relative to $\Psi^{(1)}$, $\Psi^{(2)}$. The number of distinct characters in an alphabet $k = 4, 8, 12, 16, 20, 24$

n	k					
	4	8	12	16	20	24
20	$_{6}9.29_{13}$	$_{9}12.12_{15}$	$_{10}13.57_{16}$	$_{12}14.46_{19}$	$_{12}15.10_{19}$	$_{12}15.67_{18}$
25	$_{7}10.59_{14}$	$_{10}14.30_{17}$	$_{12}16.33_{20}$	$_{15}17.77_{21}$	$_{15}18.46_{21}$	$_{15}18.97_{21}$
30	$_{7}12.05_{16}$	$_{13}16.89_{21}$	$_{14}19.04_{23}$	$_{17}20.74_{23}$	$_{18}21.50_{25}$	$_{19}22.52_{26}$
35	$_{9}13.87_{18}$	$_{16}19.35_{22}$	$_{19}21.76_{25}$	$_{20}23.49_{27}$	$_{22}24.93_{28}$	$_{21}25.79_{29}$
40	$_{11}15.88_{19}$	$_{19}21.66_{27}$	$_{21}24.94_{28}$	$_{23}26.77_{31}$	$_{24}28.23_{31}$	$_{26}28.89_{33}$
45	$_{11}17.46_{21}$	$_{19}24.05_{28}$	$_{23}27.46_{31}$	$_{25}29.63_{33}$	$_{28}31.21_{35}$	$_{27}32.23_{36}$
50	$_{15}19.18_{23}$	$_{22}26.52_{31}$	$_{26}30.41_{34}$	$_{29}32.96_{37}$	$_{31}34.39_{38}$	$_{33}35.99_{39}$

Table 6.4 Number of cases with totally optimal sequence alignments relative to $\Psi^{(1)}$ and $\Psi^{(2)}$ among 100 trials. The number of distinct characters in an alphabet $k = 4, 8, 12, 16, 20, 24$

n	k					
	4	8	12	16	20	24
20	0	0	0	0	0	0
25	0	0	0	0	0	0
30	0	0	0	0	0	0
35	0	0	0	0	0	0
40	0	0	0	0	0	0
45	0	0	0	0	0	0
50	0	0	0	0	0	0

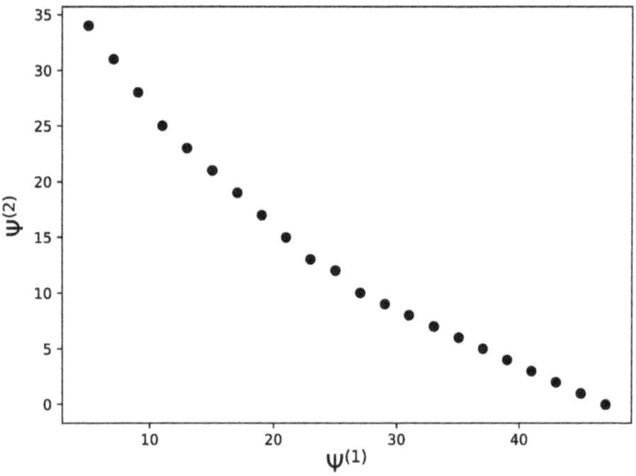

Fig. 6.4 Set of Pareto optimal points for a bi-criteria optimization of global sequence alignments relative to $\Psi^{(1)}$ and $\Psi^{(2)}$ for $n = 50$ and $k = 4$

References

1. Chikalov, I., Hussain, S., Moshkov, M., Odat, E.: Sequential optimization of global sequence alignments relative to different cost functions. In: ACM International Conference on Convergence and Hybrid Information Technology, ICHIT 2010, Daejeon, Korea, August 26–28, 2010. ACM (2010)
2. Higgins, D.: Multiple sequence alignment. In: Bishop, M.J. (ed.) Genetic Databases, Biological Techniques Series, pp. 165–183. Academic Press (1997)
3. Kleinberg, J., Tardos, E.: Algorithm Design. Addison-Wesley Longman Publishing Co., Inc. (2005)
4. Levenshtein, V.I.: Binary codes capable of correcting deletions, insertions, and reversals. Sov. Phys. Dokl. **10**(8), 707–710 (1966)
5. Needleman, S.B., Wunsch, C.D.: A general method applicable to the search for similarities in the amino acid sequence of two proteins. J. Mol. Biol. **48**(3), 443–453 (1970)
6. Smith, T., Waterman, M.: Identification of common molecular subsequences. J. Mol. Biol. **147**(1), 195–197 (1981)

Chapter 7
Optimal Paths in Directed Graphs

The problem of path length minimization in graphs is very well known. There are algorithms solving this problem based on a greedy approach such as Dijkstra's algorithm [2], algorithms based on a dynamic programming approach such as Floyd-Warshall algorithm [3, 6], etc.

In [1], an algorithm for multi-stage optimization of paths in a directed graph G relative to two types of cost functions was presented. Let w_1, w_2 be weight functions assigning nonnegative real weights to edges of G, and p be a directed path in G. Then the cost function ψ_{w_1} of the first type is equal to the sum of weights of edges in p and the cost function φ_{w_2} of the second type is equal to the minimum weight of an edge in p. The cost functions of the first type should be minimized, and the cost functions of the second type should be maximized.

Two interpretations of the graph G were considered in [1]. The first interpretation of the graph G is a road network. The nodes represent intersections and edges represent roads between the intersections. The weight can be seen as the length or pricing of a road, then ψ_{w_1} is the total length or cost of the path p. Moreover, if weight represents a horizontal or vertical clearance, then φ_{w_2} defines the maximum legal vehicle dimension in the path p. The second interpretation of the graph G is a computer network. The cost functions ψ_{w_1} of the first type can be used for the minimization of the delay or packet loss rate, where the cost functions φ_{w_2} of the second type can be used for the maximization of the bit rate.

In this chapter, we design a conventional circuit without repetitions describing the set of paths under consideration and define cost functions of two types for this circuit. We evaluate the number of operations and the time required by the algorithms for multi-stage and bi-criteria optimization of paths relative to cost functions of two types, and for counting the paths. We consider an example and discuss the results of experiments with randomly generated graphs and weight functions. This chapter is partially based on [5].

M. Mankowski and M. Moshkov, *Dynamic Programming Multi-Objective Combinatorial Optimization*, Studies in Systems, Decision and Control 331,
https://doi.org/10.1007/978-3-030-63920-4_7

7.1 Definition of Circuit and Cost Functions

Let $G = (V, E)$ be a directed graph without loops and multiple edges, where V is the set of nodes and E is the set of edges of G. We denote $n = |V|$ and $m = |E|$, and assume that $n \geq 2$ and $m \geq n - 1$. The graph G is given by its adjacency list.

The numbers n and m are the dimensions of problems under consideration related to the path optimization in graph G. To define a cost function for the path optimization problem, we need to give m weights of edges. To describe an instance of the problem, we need the adjacency list and one or two cost functions. Note that m is at most the number of numeric parameters of the problem, and n is at most the number of numeric parameters plus one.

If weights are nonnegative integers, then the adjacency list and the binary representations of m or $2m$ numbers defining one or two cost functions form the representation of the problem instance (we will omit the word instance). It is clear that m and n are at most the length of the problem representation.

A directed path in G is represented by a sequence of nodes $u_1, \ldots, u_k \in V$ such that, for $i = 1, \ldots, k - 1$, $(u_i, u_{i+1}) \in E$.

Let s, t be two nodes in G such that there exists a directed path from s to t. We call s the source node, and t the target node.

We consider the problem of finding optimal paths in the graph G from s to t. The optimization criteria are defined by cost functions.

We study two types of cost functions. Each cost function assigns a cost to each directed path in G. Let w_t, $t = 1, 2$, be a weight function that assigns to each edge (u_i, u_j) of G a weight $w_t(u_i, u_j)$, which is a nonnegative real number.

Let $p = u_1, u_2, \ldots, u_k, u_{k+1}$ be a directed path in G. A cost function of the first type ψ_{w_1} is defined as follows: $\psi_{w_1}(p) = \sum_{i=1}^{k} w_1(u_i, u_{i+1})$. If $k = 0$, then $\psi_{w_1}(p) = 0$. We aim to minimize the value of this function. A cost function of the second type φ_{w_2} is defined as follows: $\varphi_{w_2}(p) = \min\{w_2(u_1, u_2), \ldots, w_2(u_k, u_{k+1})\}$. If $k = 0$, then $\varphi_{w_2}(p) = +\infty$. We aim to maximize the value of this function.

Let p be a directed path in G from s to t. One can show that there exists a directed path p' from s to t in G, which can be obtained from p by the removal of some nodes and edges such that it satisfies the following conditions:

(a) The length of p' is at most $|V| - 1$,
(b) Both s and t appear exactly once in p',
(c) $\psi_{w_1}(p') \leq \psi_{w_1}(p)$,
(d) $\varphi_{w_2}(p') \geq \varphi_{w_2}(p)$.

Therefore, later on, we will focus on the study of paths in G from s to t satisfying the conditions (a) and (b). We denote by $P(G)$ the set of all such paths.

We introduce a directed acyclic graph $\Gamma_0 = \Gamma_0(G)$, which represents the set of directed paths $P(G)$ from s to t satisfying conditions (a) and (b).

Let $V = \{v_1, \ldots, v_n\}$, where $v_1 = s$ and $v_n = t$. The set of nodes of Γ_0 is divided into n layers. The first layer contains the only node v_1^1. The last layer contains the only node v_n^n. For $k = 2, \ldots, n-1$, the kth layer contains $n-2$ nodes $v_2^k, \ldots, v_{n-1}^k$. In total, we have at most n^2 nodes in Γ_0.

Let $i, j \in \{2, \ldots, n-1\}$. Then we have an edge from v_1^1 to v_j^2 if and only if $(v_1, v_j) \in E$. For $k = 2, \ldots, n-2$, we have an edge from v_i^k to v_j^{k+1} if and only if $(v_i, v_j) \in E$.

Let $2 \le j \le n-1$ and $2 \le k \le n-1$. Then we have an edge from v_j^k to v_n^n if and only if $(v_j, v_n) \in E$. We have an edge from v_1^1 to v_n^n if and only if $(v_1, v_n) \in E$. There are no other edges in Γ_0. In total, we have at most $m(n-1) \le mn$ edges in Γ_0.

We denote by $P(\Gamma_0)$ the set of all directed paths in Γ_0 from v_1^1 to v_n^n. It is clear that there is a one-to-one mapping from the set $P(G)$ to the set $P(\Gamma_0)$. The path $p = v_1, v_{j_2}, \ldots, v_{j_t}, v_n$ from $P(G)$ corresponds to the path $\tilde{p} = v_1^1, v_{j_2}^2, \ldots, v_{j_t}^t, v_n^n$ from $P(\Gamma_0)$.

We denote by Γ the graph obtained from Γ_0 by the removal of all nodes that do not belong to any directed path from v_1^1 to v_n^n. We also remove all edges that are incident to the removed nodes. We denote by $P(\Gamma)$ the set of all directed paths in Γ from v_1^1 to v_n^n. It is clear that $P(\Gamma) = P(\Gamma_0)$.

For a node u from Γ, let $P(u)$ be the set of all directed paths from v_1^1 to u. Let (u_1, u_2) be an edge from Γ. We denote by $P(u_1, u_2)$ the set $\{\mu_{(u_1,u_2)}(p) : p \in P(u_1)\}$, where $\mu_{(u_1,u_2)}$ is a concatenation function defined on $P(u_1)$, which corresponds to each path $p \in P(u_1)$ the path $\mu_{(u_1,u_2)}(p) = p, u_2$.

We have $P(v_1^1) = \{v_1^1\}$, where v_1^1 is the path of length 0. Let u be a node of Γ with k entering edges $e_1, \ldots, e_k$ leaving the nodes $u_1, \ldots, u_k$, $k > 0$. Then

$$P(u) = \bigcup_{i=1}^{k} P(u_i, u) .$$

Let $U = \bigcup_{v \in V(\Gamma)} P(u)$, where $V(\Gamma)$ is the set of nodes of Γ. We now transform the graph Γ into a circuit S without repetitions over the set U, which describes the set $P(\Gamma)$ of all directed paths from v_1^1 to v_n^n in the graph Γ.

We transform the source node v_1^1 of Γ into the only input node v_1^1 of S, which is labeled with the set $\{v_1^1\}$. Despite node v_1^1, each remaining node v_i^k of Γ becomes a unifying node v_i^k of the circuit S labeled with the symbol $\cup$. Note that for each node of Γ, the corresponding node in S has the same notation.

We transform each edge (u_1, u_2) from Γ into a functional node $v_{(u_1,u_2)}$ of S labeled with the concatenation function $\mu_{(u_1,u_2)}$ that has one entering edge leaving u_1 (we will not label this edge with the number 1) and one leaving edge entering u_2.

The node v_n^n is the output node for the circuit S and is labeled with $*$.

The circuit S contains one input node, at most n^2 unifying nodes, and at most mn functional nodes. Each functional node has exactly one entering edge. Each unifying node with the exception of v_n^n has at most m entering edges. The unifying node v_n^n has at most mn entering edges. Hence, the circuit S contains at most $n^2 + mn + 1$ nodes and at most $n^2m + 2mn$ edges. It is easy to check that the circuit S is reachable (see Remark 3.4).

Let u be a node of Γ. One can show that $S(u) = P(u)$, where $P(u)$ is the set of paths corresponding to the node u of Γ and $S(u)$ is the set of paths corresponding to the node u in S. Let (u_1, u_2) be an edge in Γ. One can show that $S(v_{(u_1,u_2)}) = P(u_1, u_2)$.

We now prove that S is a circuit without repetitions. It is clear that the concatenation function $\mu_{(u_1,u_2)}$ is injective for each edge (u_1, u_2) in Γ. Let u_1 and u_2 be two different nodes from $V_i(S) \cup V_f(S)$. We now show that $S(u_1) \cap S(u_2) = \emptyset$. Let u_1 be the input node and u_2 be a functional node. Then $S(u_1)$ contains the only path of the length zero, but $S(u_2)$ does not contain any path of the length zero. Let u_1 and u_2 be two functional nodes corresponding to different edges e_1 and e_2 in Γ. Then the last edge in each path from $S(u_i)$ is e_i, $i = 1, 2$. Therefore, $S(u_1) \cap S(u_2) = \emptyset$.

Let w_t, $t = 1, 2$, be a weight function that assigns to each edge (v_i, v_j) of G a weight $w_t(v_i, v_j)$, which is a nonnegative real number. These weight functions are extended to Γ in a natural way: $w_t(v_i^{k_1}, v_j^{k_2}) = w_t(v_i, v_j)$ for $t = 1, 2$ if $(v_i^{k_1}, v_j^{k_2})$ is an edge in Γ.

We now define two cost functions Ψ^{w_1} and Φ^{w_2} for the circuit S given by functions ψ^{w_1}, $\psi_u^{w_1}$ and φ^{w_2}, $\varphi_u^{w_2}$ for each $u \in V_f(S)$, respectively. Let $\psi^{w_1}(v_1^1) = 0$ and $\varphi^{w_2}(v_1^1) = -\infty$. Let (u_1, u_2) be an edge in Γ. Then $\psi^{w_1}_{v_{(u_1,u_2)}}(x) = x + w_1(u_1, u_2)$ and $\varphi^{w_2}_{v_{(u_1,u_2)}}(x) = \max\{x, -w_2(u_1, u_2)\}$.

These cost functions are adapted (see Sect. 2.4). They are described by formulas over the set of elementary operations $B = \{x + y, \max(x, -y)\}$. Note that to compute the value of ψ^{w_1} or φ^{w_2}, we do not need any operations and to compute the value of $\psi^{w_1}_{v_{(u_1,u_2)}}$ or $\varphi^{w_2}_{v_{(u_1,u_2)}}$, we need only one operation from B.

It is clear that Ψ^{w_1} is a strictly increasing cost function and Φ^{w_2} is an increasing cost function. If values of w_1 and w_2 are nonnegative integers, then Ψ^{w_1} is a strictly increasing nonnegative integer cost function for S and Φ^{w_2} is an increasing nonpositive integer cost function for S.

Let $p \in P(G)$ and $\tilde{p}$ be the path in $P(\Gamma)$ corresponding to p. One can show that $\Psi^{w_1}(\tilde{p}) = \psi_{w_1}(p)$ and $\Phi^{w_2}(\tilde{p}) = -\varphi_{w_2}(p)$.

We have no tools for the maximization of cost functions. So, instead of maximization of φ_{w_2}, we will minimize Φ^{w_2}. Later on, in algorithms of multi-stage and bi-criteria optimization, we will use only cost functions of the kind Ψ^{w_1} or Φ^{w_2}.

7.2 Time Complexity of Algorithms

We assume that the directed graph $G = (V, E)$ without loops and multiple edges, and with $n \geq 2$ nodes and $m \geq n - 1$ edges is given by its adjacency list. The graph Γ_0 with at most n^2 nodes and at most mn edges can be obtained from G in $O(n^2 + mn)$ time. Using breadth-first search two times (from the node v_1^1 in the graph Γ_0 and from the node v_n^n in the graph Γ_0 with the reverse direction of the edges), we can find in the graph Γ_0 all nodes that do not belong to any directed path from v_1^1 to v_n^n and construct the graph Γ in $O(n^2 + mn)$ time. The circuit S is fully described by the graph Γ. It can be constructed from the graph Γ in time $O(n^2m)$. The total time to construct the scheme of the circuit S is $O(n^2m)$.

The algorithms $\mathcal{A}_1$, $\mathcal{A}_2$, and $\mathcal{A}_4$ work with the scheme of the circuit S (see Sect. 2.4). Note that the algorithms $\mathcal{A}_1$ and $\mathcal{A}_2$ can also work with the schemes of edge-preserving subcircuits of S obtained as a result of the algorithm $\mathcal{A}_1$ work—see Remark 3.1.

Denote $B^+ = B \cup \{x : y\}$, where $x : y$ is the operation of comparison with values $x < y$, $x = y$, and $x > y$.

Let w be a weight function that has nonnegative integer values. Denote $\nu_w = \max\{w(v_i, v_j) : (v_i, v_j) \in E\}$. Then $N_{\Psi^w}(S) \leq \nu_w(n-1)$ and $N_{\Phi^w}(S) \leq \nu_w$, where $N_{\Psi^w}(S) = \max\{|\Psi^w(\delta)| : \delta \in U(S)\}$, $N_{\Phi^w}(S) = \max\{|\Phi^w(\delta)| : \delta \in U(S)\}$, and $U(S) = \bigcup_{v \in V(S)} S(v)$. It is clear that $\log_2 \nu_w$ is at most the length of the problem representation if the problem uses at least one of the cost functions Ψ^w, Φ^w.

We can use the algorithm $\mathcal{A}_1$ for multi-stage optimization of paths. Using Theorem 3.1, we evaluate the number of operations from B^+ that the algorithm $\mathcal{A}_1$ makes while working with the scheme of an edge-preserving subcircuit T of the circuit S and a cost function Θ for S of the kind Ψ^w or Φ^w.

The scheme of T contains one input node. We do not need any operations to process this node. The scheme of T contains at most n^2 unifying nodes. The unifying node v_n^n has at most mn entering edges, where each of the remaining unifying nodes has at most m entering edges. Then, the algorithm $\mathcal{A}_1$ makes at most $(2m-1)n^2 + 2mn - 1 \leq 4mn^2$ comparisons. The scheme contains at most mn functional nodes. For each functional node, the algorithm $\mathcal{A}_1$ makes one operation from B^+. To process all functional nodes, it requires at most $mn \leq mn^2$ operations. The total number of operations from the set B^+ made by $\mathcal{A}_1$ is at most $4mn^2 + mn^2 \leq 5mn^2$, i.e., it is polynomial depending on the problem dimensions n and m.

In the software model of computation, for the considered scheme of the circuit and cost function, the algorithm $\mathcal{A}_1$ has polynomial time complexity $O(mn^2)$ depending on the problem dimensions n and m.

Let the weight function w specifying the cost function Θ have nonnegative integer values. Using Remark 3.2, we obtain that in the integer model of computation, for the considered scheme of the circuit and cost function, the algorithm $\mathcal{A}_1$ has time complexity $O(mn^2t)$, where $t = \log_2(\nu_w n)$ if $\Theta = \Psi^w$ and $t = \log_2 \nu_w$ if $\Theta = \Phi^w$, which is polynomial depending on the length of the problem representation (n, m, and $\log_2 \nu_w$ are at most the length of the problem representation).

To count all possible paths in the graph described by T, we apply the algorithm $\mathcal{A}_2$. We use Theorem 3.2 to evaluate the number of operations from B^+ made by $\mathcal{A}_2$. Despite the unifying node v_n^n with at most mn entering edges, each of the remaining unifying nodes has at most m entering edges. Therefore, we need at most $(m-1)n^2 + mn - 1 \leq 2mn^2$ additions to process all unifying nodes. To treat all functional nodes and the input node, we do not need any operations. The total number of operations from B^+ (additions) is at most $2mn^2$, i.e., it is polynomial depending on the problem dimensions n and m.

In the software model of computation, for the considered scheme of the circuit, the algorithm $\mathcal{A}_2$ has polynomial time complexity $O(mn^2)$ depending on the problem dimensions n and m.

We now evaluate the number $|S(v_n^n)|$, which is equal to the number of directed paths from v_1^1 to v_n^n in the graph Γ. This number is equal to the number of paths in the set $P(G)$. Each path from $P(G)$ is a sequence of edges $e_1, \ldots, e_{k-1}, e_k$ from E such that $k < n$. We correspond to this path the tuple $(e_1, \ldots, e_{k-1}, e_k, \ldots, e_k)$ from E^n. Since G does not contain loops, tuples corresponding to different paths from $P(G)$ are different. Therefore, the set $P(G)$ contains at most m^n paths. We know that the circuit S is reachable. Using Remark 3.4, we obtain that in the integer model of computation, for the considered scheme of the circuit, the algorithm $\mathcal{A}_2$ has the time complexity $O(mn^3 \log m)$, which is polynomial depending on the length of the problem representation.

Let us consider two cost functions $\Theta^{(1)}, \Theta^{(2)}$ specified by the weight functions w_1 and w_2, respectively. We can use the algorithm $\mathcal{A}_4$ to find the set of Pareto optimal points for the problem of bi-criteria optimization of paths relative to $\Theta^{(1)}$ and $\Theta^{(2)}$.

We denote $M = M_{\Theta^{(1)},\Theta^{(2)}}(S)$, where

$$M_{\Theta^{(1)},\Theta^{(2)}}(S) = \max\{|Par(t_{\Theta^{(1)},\Theta^{(2)}}(S, v))| : v \in V(S)\} .$$

To evaluate the number of operations of the algorithm $\mathcal{A}_4$, we use Theorem 4.1. To treat the input node of the scheme of S, we do not need any operations.

The scheme of S contains at most mn functional nodes. Each node has one entering edge. The algorithm needs at most $2mnM \le 2mn^2M\log_2(2M)$ operations from B and at most $4mnM\log_2 M \le 4mn^2M\log_2(2M)$ comparisons to process all functional nodes. In total, it needs at most $6mn^2M\log_2(2mnM)$ operations from B^+.

For the node v_n^n with at most mn entering edges, the algorithm $\mathcal{A}_4$ makes at most $4mnM\log_2(mnM)$ comparisons. For each of remaining at most n^2 unifying nodes with at most m entering edges, it makes at most $4mM\log_2(mM)$ comparisons. Then, to process all unifying nodes, the algorithm needs at most $4mnM\log_2(mnM) + 4mn^2M\log_2(mM) \le 8mn^2M\log_2(2mnM)$ comparisons.

Hence, the algorithm $\mathcal{A}_4$ makes at most $14mn^2M\log_2(2mnM)$ operations.

Let $\Theta^{(1)} = \Psi^{w_1}$ and w_1 have nonnegative integer values. Then $N_{\Psi^{w_1}}(S) \le \nu_{w_1}(n-1)$. By Proposition 4.4, $M_{\Theta^{(1)},\Theta^{(2)}}(S) \le \nu_{w_1}(n-1)+1$. Since $M \le \nu_{w_1}(n-1)+1$, the algorithm $\mathcal{A}_4$ makes at most

$$\begin{aligned} &14mn^2(\nu_{w_1}(n-1)+1)\log_2(2mn(\nu_{w_1}(n-1)+1)) \\ &\qquad \le 28mn^3(\nu_{w_1}+1)\log_2(2mn(\nu_{w_1}+1)) \end{aligned}$$

operations from B^+, i.e., pseudo-polynomial number of operations (polynomial depending on the dimensions n and m, and the numeric parameter ν_{w_1}).

Let $\Theta^{(1)} = \Phi^{w_1}$. It is clear that $|\{\Phi(\delta) : \delta \in U(S)\}| \le m$. By Proposition 4.5, $M_{\Theta^{(1)},\Theta^{(2)}}(S) \le m$. Since $M \le m$, the algorithm $\mathcal{A}_4$ makes at most

$$14m^2n^2\log_2(2m^2n)$$

operations from B^+, i.e., polynomial number of operations depending on the dimensions n and m.

In the software model of computation, for the considered circuit and pair of cost functions, the algorithm $\mathcal{A}_4$ has the time complexity $O(mn^3\nu_{w_1}\log(mn\nu_{w_1}))$ if $\Theta^{(1)} = \Psi^{w_1}$, which is pseudo-polynomial (polynomial depending on the dimensions n and m, and the numeric parameter ν_{w_1}), and $O(m^2n^2\log(mn))$ if $\Theta^{(1)} = \Phi^{w_1}$, which is polynomial depending on the dimensions n and m.

Let us assume now that w_1 and w_2 have nonnegative integer values. Using Remark 4.4, we obtain that in the integer model of computation, for the considered circuit and pair of cost functions, the algorithm $\mathcal{A}_4$ has the time complexity

$$O(mn^3\nu_{w_1}\log(mn\nu_{w_1})\log(\max(\nu_{w_1}, \nu_{w_2})n))$$

if $\Theta^{(1)} = \Psi^{w_1}$, which is pseudo-polynomial (polynomial depending on the length of the problem representation and the numeric parameter ν_{w_1}), and

$$O(m^2n^2\log(mn)\log(\max(\nu_{w_1}, \nu_{w_2})n))$$

if $\Theta^{(1)} = \Phi^{w_1}$, which is polynomial depending on the length of the problem representation.

Algorithms from [4] mentioned in Sect. 1.3 require less number of operations. If $\Theta^{(1)} = \Phi^{w_1}$, then the required number of operations is $O(m^2\log n)$. Let $\Theta^{(1)} = \Psi^{w_1}$, $\Theta^{(2)} = \Psi^{w_2}$ and $\nu_{w_1} \leq \nu_{w_2}$. Then the required number of operations is $O(nm\nu_{w_1}\log(n\nu_{w_1}))$.

7.3 Example

We consider a directed graph $G = (V, E)$ (as depicted in Fig. 7.1), where $s = v_1$ and $t = v_4$, and two weight functions w_1 and w_2 for this graph such as

$$w_1(e) = 1 \quad \text{for all } e \in E ,$$

and

$$w_2(v_1, v_2) = 1,\ w_2(v_2, v_4) = 4,\ w_2(v_1, v_3) = 3 ,$$
$$w_2(v_3, v_4) = 0,\ w_2(v_2, v_3) = 4,\ w_2(v_3, v_2) = 5 .$$

Initially, we create the graph Γ_0 (see Fig. 7.2), note that $\Gamma_0 = \Gamma$. Then we transform Γ into a circuit S_{OP} (see Fig. 7.3).

Fig. 7.1 Graph G

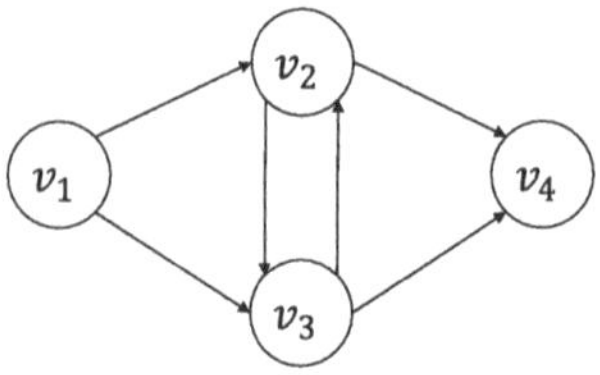

Fig. 7.2 Directed acyclic graph $\Gamma_0 = \Gamma$

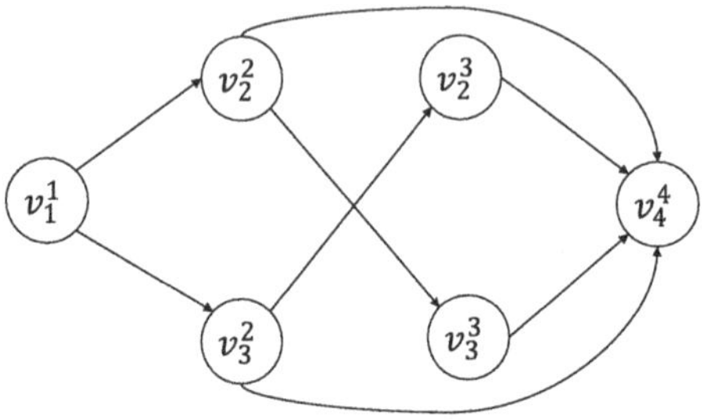

Fig. 7.3 Circuit S_{OP}

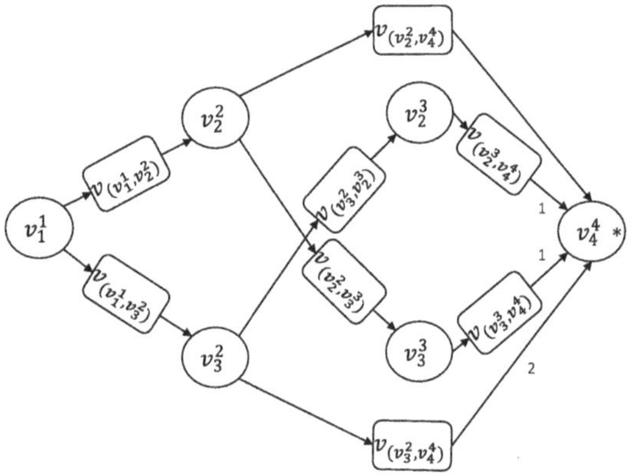

We apply the procedure of optimization relative to Ψ^{w_1} to the circuit S_{OP}. As a result, we have the subcircuit $S_{OP}^{\Psi^{w_1}}$, which can be obtained from S_{OP} by the removal of edges labeled with 1 (see Fig. 7.3). The set $P(S_{OP}^{\Psi^{w_1}})$ contains exactly two paths.

We apply the procedure of optimization relative to Φ^{w_2} to the circuit $S_{OP}^{\Psi^{w_1}}$. As a result, we have the circuit $S_{OP}^{\Psi^{w_1},\Phi^{w_2}}$, which can be obtained from S_{OP} by the removal of edges labeled with 1 and 2 (see Fig. 7.3). The set $P(S_{OP}^{\Psi^{w_1},\Phi^{w_2}})$ contains exactly one path.

We apply the procedure of bi-criteria optimization relative to Ψ^{w_1} and Φ^{w_2} to the circuit S_{OP}. The set of Pareto optimal points for each node of S_{OP} is depicted in Fig. 7.4.

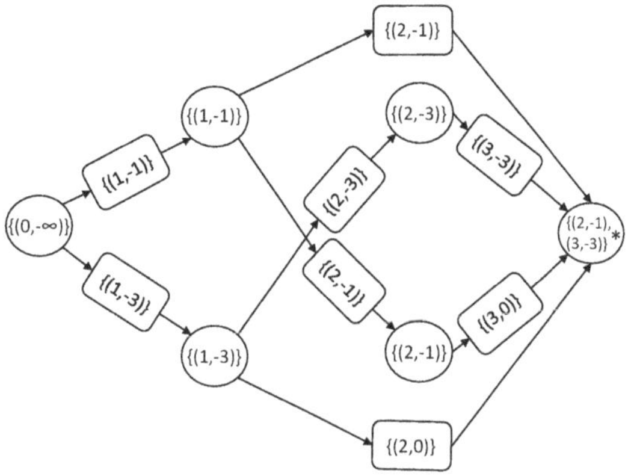

Fig. 7.4 Circuit S_{OP} with the sets of Pareto optimal points associated with each node

7.4 Experiments

We experimentally tested different optimization scenarios for paths in directed graphs. Each experiment was repeated ten times.

Let $\alpha \in \{0.2, 0.4, 0.6, 0.8\}$. For a given n, we generated directed graphs without loops and multiple edges that contain n nodes $s = v_1, \ldots, v_n = t$. We randomly generated a directed path from s to t and after that added randomly new edges until the total number of edges is equal to $\lceil \alpha n^2 \rceil$. For each edge, the integer values of weight functions w_1 and w_2 were randomly generated with uniform distribution from the interval [0, 100]. Denote $\Psi = \Psi^{w_1}$ and $\Phi = \Phi^{w_2}$.

Table 7.1 shows, for $n = 50, 60, \ldots, 120$, the average number of directed paths described by the circuit before optimization, after optimization relative to Ψ, relative to Φ, relative to Ψ, Φ, and relative to Φ, Ψ. Graphs had n nodes and $\lceil 0.2n^2 \rceil$ edges in A or $\lceil 0.8n^2 \rceil$ edges in B. Experiments resulted in many paths optimal relative to Φ but in a few relative to Ψ. For all tested scenarios there were at most 1.1 (on average) paths after the optimization relative to Ψ, Φ in A and B. The same situation is with the optimization relative to Φ, Ψ.

Table 7.2 shows the average value of Ψ after optimization relative to Ψ and Φ, Ψ, and the average value of Φ after optimization relative to Φ and Ψ, Φ. Graphs had n nodes and $\lceil 0.2n^2 \rceil$ edges in A or $\lceil 0.8n^2 \rceil$ edges in B. With the increase in the number of edges (we compare A and B), the values of Ψ and Φ decreased. For both cost functions in A and B, the two-stage optimization impacted the value of the final stage cost function. For instance, the value of Ψ in B increased from 6.8 to 187.4 for $n = 120$ comparing optimizations relative to Ψ and Φ, Ψ.

We also consider bi-criteria optimization of paths. Table 7.3 shows the minimum, maximum, and the average number of Pareto optimal points for bi-criteria optimization problem relative to Ψ, Φ for graphs with $n = 50, 60, \ldots, 120$ nodes and $\lceil \alpha n^2 \rceil$,

Table 7.1 Average number of directed paths described by the circuit before optimization, after optimization relative to Ψ, relative to Φ, relative to Ψ, Φ, and relative to Φ, Ψ among 10 trials for graphs with n nodes and $\lceil \alpha n^2 \rceil$ edges, where $\alpha = 0.2$ for A and $\alpha = 0.8$ for B

A					
n	$\emptyset$	Ψ	Φ	Ψ, Φ	Φ, Ψ
50	9.17×10^{47}	1.00	3.35×10^{15}	1.00	1.10
60	7.76×10^{61}	1.10	2.26×10^{12}	1.00	1.00
70	1.25×10^{77}	1.20	9.65×10^{28}	1.00	1.00
80	2.18×10^{93}	1.00	4.72×10^{49}	1.00	1.00
90	6.26×10^{109}	1.10	1.92×10^{34}	1.00	1.00
100	1.29×10^{127}	1.20	1.42×10^{58}	1.00	1.00
110	2.33×10^{144}	1.10	1.79×10^{80}	1.10	1.00
120	1.31×10^{162}	1.20	2.03×10^{98}	1.10	1.00
B					
n	$\emptyset$	Ψ	Φ	Ψ, Φ	Φ, Ψ
50	6.00×10^{76}	1.20	3.22×10^{14}	1.00	1.00
60	2.82×10^{97}	1.40	1.21×10^{24}	1.00	1.10
70	5.75×10^{118}	1.40	1.11×10^{27}	1.10	1.00
80	5.86×10^{140}	1.20	2.46×10^{33}	1.00	1.00
90	2.12×10^{163}	1.30	9.04×10^{46}	1.00	1.00
100	2.63×10^{186}	1.20	1.30×10^{44}	1.00	1.00
110	6.92×10^{209}	1.30	3.91×10^{57}	1.10	1.00
120	6.23×10^{233}	1.10	3.05×10^{87}	1.00	1.00

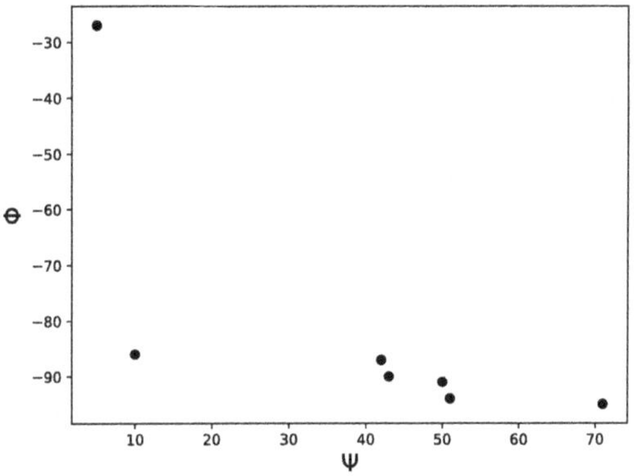

Fig. 7.5 Set of Pareto optimal points for a bi-criteria optimization of paths relative to Ψ and Φ, $n = 120$, $\alpha = 0.8$

$\alpha = 0.2, 0.4, 0.6, 0.8$, edges. There were at most 22 Pareto optimal points. There were no totally optimal paths relative to Ψ and Φ (see Table 7.4). The set of Pareto optimal points for a bi-criteria optimization of paths relative to Ψ and Φ for $n = 120$ and $\alpha = 0.8$ is depicted in Fig. 7.5.

Table 7.2 Average value of Ψ after optimization relative to Ψ and Φ, Ψ, and average value of Φ after optimization relative to Φ and Ψ, Φ among 10 trials for graphs with n nodes and $\lceil \alpha n^2 \rceil$ edges, where $\alpha = 0.2$ for A and $\alpha = 0.8$ for B

A

n	Ψ	Φ, Ψ	Φ	Ψ, Φ
50	47.40	227.30	−82.20	−33.70
60	39.20	256.80	−89.70	−15.40
70	30.30	302.00	−83.90	−20.50
80	32.30	311.40	−88.00	−20.50
90	29.60	388.80	−89.60	−24.40
100	26.10	400.60	−89.90	−30.60
110	32.60	361.10	−89.50	−31.50
120	25.90	337.70	−91.70	−22.40

B

n	Ψ	Φ, Ψ	Φ	Ψ, Φ
50	11.40	243.60	−96.10	−34.30
60	9.50	289.70	−97.30	−29.10
70	10.30	262.10	−96.80	−23.60
80	8.80	274.30	−97.60	−32.40
90	8.20	301.10	−97.40	−20.10
100	7.20	341.80	−98.40	−44.60
110	8.80	239.20	−97.80	−16.00
120	6.80	187.40	−98.10	−33.40

Table 7.3 Number of Pareto optimal points (in format $_{min}average_{max}$ among 10 trials) for bi-criteria optimization problem relative to Ψ, Φ. Graphs with n nodes and $\lceil \alpha n^2 \rceil$, $\alpha = 0.2, 0.4, 0.6, 0.8$, edges

n	α			
	0.2	0.4	0.6	0.8
50	$_{2}5.60_{11}$	$_{7}9.40_{13}$	$_{5}9.60_{19}$	$_{5}10.00_{18}$
60	$_{4}8.60_{12}$	$_{4}7.80_{14}$	$_{7}10.30_{13}$	$_{5}8.20_{11}$
70	$_{4}8.80_{12}$	$_{7}10.70_{16}$	$_{4}9.20_{14}$	$_{7}9.40_{14}$
80	$_{4}7.40_{11}$	$_{6}9.70_{14}$	$_{3}9.60_{15}$	$_{5}10.80_{18}$
90	$_{4}9.10_{13}$	$_{7}11.60_{17}$	$_{9}12.80_{19}$	$_{4}9.40_{14}$
100	$_{6}9.50_{16}$	$_{2}10.00_{16}$	$_{9}10.80_{13}$	$_{8}11.50_{19}$
110	$_{3}7.80_{14}$	$_{8}12.20_{16}$	$_{7}13.60_{22}$	$_{9}12.30_{16}$
120	$_{7}10.40_{18}$	$_{8}12.70_{20}$	$_{11}15.50_{20}$	$_{7}11.70_{14}$

Table 7.4 Number of cases with totally optimal paths relative to Ψ and Φ among 10 trials. Graphs with n nodes and $\lceil \alpha n^2 \rceil$, $\alpha = 0.2, 0.4, 0.6, 0.8$, edges

n	α			
	0.2	0.4	0.6	0.8
50	0	0	0	0
60	0	0	0	0
70	0	0	0	0
80	0	0	0	0
90	0	0	0	0
100	0	0	0	0
110	0	0	0	0
120	0	0	0	0

References

1. AbuBekr, J., Chikalov, I., Hussain, S., Moshkov, M.: Sequential optimization of paths in directed graphs relative to different cost functions. In: Sato, M., Matsuoka, S., Sloot, P.M.A., van Albada, G.D., Dongarra, J. (eds.) International Conference on Computational Science, ICCS 2011, Nanyang Technological University, Singapore, June 1–3, 2011, Procedia Comput. Sci., vol. 4, pp. 1272–1277. Elsevier (2011)
2. Dijkstra, E.W.: A note on two problems in connexion with graphs. Numer. Math. **1**(1), 269–271 (1959)
3. Floyd, R.W.: Algorithm 97: Shortest path. Commun. ACM **5**(6), 345 (1962)
4. Hansen, P.: Bicriterion path problems. In: Fandel, G., Gal, T. (eds.) Multiple Criteria Decision Making Theory and Application. Lect. Notes Econ. Math., vol. 177, pp. 109–127. Springer (1980)
5. Mankowski, M., Moshkov, M.: Dynamic programming bi-criteria combinatorial optimization. Discret. Appl. Math. **284**, 513–533 (2020)
6. Warshall, S.: A theorem on boolean matrices. J. ACM **9**(1), 11–12 (1962)

Chapter 8
Binary Search Trees

Knuth in [4] studied the problem of average depth minimization for binary search trees with two kinds of keys related to successful and unsuccessful search with their probabilities. In [1], this problem was generalized by associating a weight to each successful key, which can correspond to a comparison cost of key and searching value or to cost of memory access, e.g., some keys are stored in secondary memory with a higher cost of access [7]. In the same paper [1], multi-stage optimization of binary search trees related to the weighted depth and average weighted depth was considered.

In this chapter, we study multi-stage and bi-criteria optimization of binary search trees using the optimization framework based on the consideration of a conventional circuit without repetitions. We construct such a circuit, define two cost functions for it, evaluate the number of operations and the time required by the algorithms, consider an example, and discuss the results of computational experiments. This chapter is partially based on [5].

8.1 Definition of Circuit and Cost Functions

First, we discuss the notion of a binary search tree described in [4] (see also [2], Sect. 15.5).

Let us consider a sequence of $n \geq 1$ distinct keys $k_1, k_2, \ldots, k_n$ assuming $k_1 < k_2 < \cdots < k_n$. We also have $n + 1$ dummy keys $d_0, d_1, \ldots, d_n$ such that, for $i = 1, 2, \ldots, n - 1$, the dummy key d_i corresponds to all values between k_i and $k_i + 1$ with d_0 representing all values less than k_1 and d_n representing all values greater than k_n.

M. Mankowski and M. Moshkov, *Dynamic Programming Multi-Objective Combinatorial Optimization*, Studies in Systems, Decision and Control 331,
https://doi.org/10.1007/978-3-030-63920-4_8

Fig. 8.1 Binary search tree $tree(d_i)$

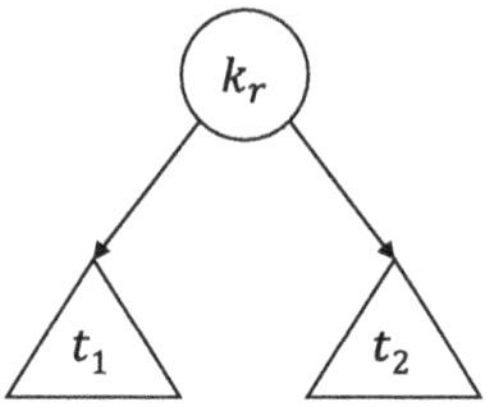

Fig. 8.2 Binary search tree $TREE(k_r, t_1, t_2)$

For any $i, j, 0 \leq i \leq j \leq n$, define the set $T(i, j)$ of all binary search trees with keys $k_{i+1}, \ldots, k_j$ and dummy keys $d_i, \ldots, d_j$. For $i = j$, the set $T(i, i)$ contains exactly one binary search tree $tree(d_i)$ depicted in Fig. 8.1. Let $i < j$ and $i < r \leq j$. We denote

$$T(i, r, j) = \{TREE(k_r, t_1, t_2) : t_1 \in T(i, r-1), t_2 \in T(r, j)\} ,$$

where the tree $TREE(k_r, t_1, t_2)$ is depicted in Fig. 8.2. Then

$$T(i, j) = \bigcup_{r=i+1}^{j} T(i, r, j) .$$

The set $T(0, n)$ is equal to the set of all binary search trees with keys $k_1, \ldots, k_n$ and dummy keys $d_0, \ldots, d_n$.

The number n is the dimension of the problems under consideration related to the optimization of binary search trees. To define a weighted depth for the binary search trees, we must specify n positive integer weights of keys. To define an average weighted depth for the binary search trees, we need to specify additionally $2n + 1$ probabilities of keys and dummy keys, which are not integers in general. To describe an instance of the problem, we need one or two cost functions. Note that n is at most the number of numeric parameters of the problem.

If we consider the optimization of binary search trees relative to the weighted depth only, then the binary representations of n weights form the representation of the problem instance (we will omit the word instance). It is clear that n is at most the length of the problem representation.

Denote $U = \bigcup_{0 \leq i \leq j \leq n} T(i, j)$. We now define a circuit S without repetitions over the set U, which describes the set $T(0, n)$.

For each $i, r, j, 0 \leq i < r \leq j \leq n$, we define a merging function $\mu_{i,r,j}$ from the set $T(i, r-1) \times T(r, j)$ to the set $T(i, j)$. Let $t_1 \in T(i, r-1)$ and $t_2 \in T(r, j)$. Then $\mu_{i,r,j}(t_1, t_2) = TREE(k_r, t_1, t_2)$.

For $i = 0, \ldots, n$, the circuit S contains the input node $v(i, i)$, which is labeled with the set $\{tree(d_i)\}$. For any $i, j, 0 \leq i < j \leq n$, the circuit S contains the unifying node $v(i, j)$ labeled with the symbol $\bigcup$. For any $i, r, j, 0 \le i < r \le j \le n$, the circuit S contains the functional node $v(i, r, j)$ labeled with the merging function $\mu_{i,r,j}$. The unifying node $v(0, n)$ is considered as the output of S and is labeled with $*$.

Each unifying node $v(i, j), 0 \le i < j \le n$, has $j - i$ entering edges leaving functional nodes $v(i, r, j)$, where $i < r \le j$. Each functional node $v(i, r, j), 0 \le i < r \le j \le n$, has two entering edges labeled with the numbers 1 and 2 (we omit these numbers in the example), which leave the nodes $v(i, r-1)$ and $v(r, j)$, respectively.

One can show that $S(v(i, r, j)) = T(i, r, j)$ for any $i, r, j, 0 \le i < r \le j \le n$, and $S(v(i, j)) = T(i, j)$ for any $i, j, 0 \le i < j \le n$.

The described circuit will be considered as the preliminary version of the circuit S. The final version of S is obtained from the preliminary version of S by the removal of all nodes such that there is no directed path from the considered node to the output. We also remove all edges that are incident to the removed nodes. Later, S without mentioning the version denotes the final version of the circuit S.

The preliminary version of the circuit S contains $n + 1$ input nodes, at most $(n + 1)^3$ functional nodes with two entering edges each, and at most $(n + 1)^2$ unifying nodes with at most n entering edges each, i.e., this circuit contains at most $n + 1 + (n + 1)^2 + (n + 1)^3$ nodes and at most $3(n + 1)^3$ edges.

We now prove that S is a circuit without repetitions. It is clear that the considered merging functions $\mu_{i,r,j}$ are injective. Let v_1 and v_2 be different nodes from $V_i(S) \cup V_f(S)$. We now show that $S(v_1) \cap S(v_2) = \emptyset$. If v_1 and v_2 are input nodes, then, evidently, $S(v_1) \cap S(v_2) = \emptyset$. Let v_1 be an input node and v_2 be a functional node. Then $S(v_1)$ contains the only tree with a single node of the depth zero, but $S(v_2)$ does not contain any tree of the depth zero. Let v_1 and v_2 be functional nodes, $v_1 = v(i_1, r_1, j_1)$ and $v_2 = v(i_2, r_2, j_2)$. If $(i_1, j_1) = (i_2, j_2)$ and $r_1 \neq r_2$, the sets $S(v_1)$ and $S(v_2)$ cannot have a common tree, since $S(v_1) = T(i_1, r_1, j_1)$, $S(v_2) = T(i_1, r_2, j_1)$, and $T(i_1, r_1, j_1) \cap T(i_1, r_2, j_1) = \emptyset$—trees from these sets have different keys attached to the roots. Let $(i_1, j_1) \neq (i_2, j_2)$. One can show that $S(v_1) \subseteq T(i_1, j_1)$, $S(v_2) \subseteq T(i_2, j_2)$ and, evidently, $T(i_1, j_1) \cap T(i_2, j_2) = \emptyset$—trees from these sets use different sets of keys.

Let keys $k_1, k_2, \ldots, k_n$ have probabilities $p_1, p_2, \ldots, p_n$, and dummy keys $d_0, d_1, \ldots, d_n$ have probabilities $q_0, q_1, \ldots, q_n$. The sum of probabilities is equal to one, i.e., $\sum_{i=1}^{n} p_i + \sum_{i=0}^{n} q_i = 1$. We associate a positive integer weight w_i to each of the keys k_i, where $i = 1, \ldots, n$.

We now define two cost functions Ψ and Φ for the circuit S. Denote

$$B = \{x + y, x - y, x \cdot y, \max(x, y)\} .$$

The cost function Ψ (the weighted depth of a binary search tree) is given by the functions ψ and ψ_v for each $v \in V_f(S)$. We have $\psi(tree(d_i)) = 0$ for $i = 0, \ldots, n$, and $\psi_{v(i,r,j)}(x, y) = w_r + \max\{x, y\}$ for any functional node $v(i, r, j)$ from $V_f(S)$. To compute $\psi_{v(i,r,j)}$, we need two operations from B.

For any $i, j, 0 \le i < j \le n$, let $\pi(i, j) = \sum_{l=i+1}^{j} p_l + \sum_{l=i}^{j} q_l$. To calculate $\pi(i, j)$, we use prefix sum sequences $P_1 = P_1(0), P_1(1), \ldots, P_1(n)$, where $P_1(0) =$

0 and, for any a, $1 \leq a \leq n$, $P_1(a) = P_1(a-1) + p_a$, and $P_2 = P_2(-1), P_2(0), \ldots, P_2(n)$, where $P_2(-1) = 0$ and, for any b, $0 \leq b \leq n$, $P_2(b) = P_2(b-1) + q_b$. To compute the sequences P_1 and P_2, we need at most $2(n+1)$ operations of addition. To derive $\pi(i, j)$ from P_1 and P_2, we need three operations of addition and subtraction $\pi(i, j) = (P_1(j) - P_1(i)) + (P_2(j) - P_2(i-1))$.

The cost function Φ (the average weighted depth of a binary search tree) is given by the functions φ and φ_v for each $v \in V_f(S)$. We have $\varphi(tree(d_i)) = 0$ for $i = 0, \ldots, n$, and $\varphi_{v(i,r,j)}(x, y) = w_r \pi(i, j) + x + y$ for any functional node $v(i, r, j)$ from $V_f(S)$. To compute $\varphi_{v(i,r,j)}$ knowing P_1 and P_2, we need six operations from B.

It is clear that Ψ is an increasing nonnegative integer cost function for S and Φ is a strictly increasing cost function for S.

The considered cost functions are adapted (see Sect. 2.4). They are described by formulas over the set of elementary operations B.

8.2 Time Complexity of Algorithms

The algorithms $\mathcal{A}_1$, $\mathcal{A}_2$, and $\mathcal{A}_4$ work with the scheme of the circuit S (see Sect. 2.4). The preliminary version of the circuit S is completely described. Therefore, the time complexity of the scheme of the preliminary version of S construction is linear depending on the number of its nodes and edges, i.e., $O(n^3)$. Using breadth-first search from the output node in the scheme of the preliminary version of S with the reverse direction of the edges, we can find in the scheme of the preliminary version of S all nodes such that there is no directed path from the considered node to the output and construct the scheme of S in $O(n^3)$ time. In total, we need $O(n^3)$ time to construct the scheme of S.

Note that the algorithms $\mathcal{A}_1$ and $\mathcal{A}_2$ can also work with the schemes of edge-preserving subcircuits of S obtained as a result of the algorithm $\mathcal{A}_1$ work—see Remark 3.1.

Denote $B^+ = B \cup \{x : y\}$, where $x : y$ is the operation of comparison with values $x < y$, $x = y$, and $x > y$.

Let $\nu = \max\{w_i : 1 \leq i \leq n\}$. One can show that $N_\Psi(S) \leq \nu n$. It is clear that $\log_2 \nu$ is at most the length of the problem representation if the problem uses the cost function Ψ.

We can use the algorithm $\mathcal{A}_1$ for multi-stage optimization of binary search trees. Using Theorem 3.1, we evaluate the number of operations that the algorithm $\mathcal{A}_1$ makes when works with the scheme of an edge-preserving subcircuit T of the circuit S and a cost function from the set $\{\Psi, \Phi\}$.

The scheme of T contains at most $n + 1$ input nodes. We do not need any operations to process these nodes.

The scheme of T contains at most $(n+1)^2$ unifying nodes. Each unifying node has at most n entering edges. Then the algorithm $\mathcal{A}_1$ makes at most $2(n+1)^3$ comparisons.

The scheme of T contains at most $(n+1)^3$ functional nodes. For each functional node v, the algorithm makes at most six operations from the set B^+ to compute ψ_v or φ_v. The number of operations to process all functional nodes is at most $6(n+1)^3$. The total number of operations from the set B^+ required by $\mathcal{A}_1$ is at most $8(n+1)^3$. Note that if we consider the cost function Φ, we also need to compute at the beginning the prefix sum sequences P_1 and P_2, which requires at most $2(n+1)$ operations of addition. The whole number of operations from the set B^+ is polynomial depending on the problem dimension n.

In the software model of computation, for the considered scheme of the circuit and cost functions, the algorithm $\mathcal{A}_1$ has polynomial time complexity $O(n^3)$ depending on the problem dimension n.

Let us consider the work of the algorithm $\mathcal{A}_1$ with the cost function Ψ. Using Remark 3.2 we obtain that, for the algorithm $\mathcal{A}_1$, the number νn is an upper bound on the absolute values of the numbers at the inputs of the operations ψ_v related to the functional nodes $v \in V_f(S)$ and at the inputs of comparisons related to the unifying nodes $v \in V_u(S)$. It is easy to check that, for the algorithm $\mathcal{A}_1$, the number νn is an upper bound on the absolute values of the numbers at the inputs of operations from B in formulas for the functions ψ_v. The algorithm $\mathcal{A}_1$ uses only additions and comparisons. Therefore, in the integer model of computation, for the considered scheme of the circuit and the cost function Ψ, the algorithm $\mathcal{A}_1$ has the time complexity $O(n^3 \log(\nu n))$, which is polynomial depending on the length of the problem representation (n and $\log_2 \nu$ are at most the length of the problem representation).

To count all possible binary search trees described by the circuit T, we apply the algorithm $\mathcal{A}_2$. We use Theorem 3.2 to evaluate the number of operations required by $\mathcal{A}_2$. The scheme of T contains at most $(n+1)^2$ unifying nodes with at most n entering edges each. Therefore, we need at most $(n+1)^3$ additions to treat all unifying nodes. For each of at most $(n+1)^3$ functional nodes with two entering edges each, we need one operation of multiplication. To treat all input nodes, we do not need any operations. The total number of operations from the set B^+ is at most $2(n+1)^3$, i.e., it is polynomial depending on the dimension n.

In the software model of computation, for the considered scheme of the circuit, the algorithm $\mathcal{A}_2$ has polynomial time complexity $O(n^3)$ depending on the problem dimension n.

We now evaluate the number $|S(v(0,n))|$, which is equal to the number of binary search trees with n keys. One can show that this value is equal to the nth Catalan number C_n (see, for example, [6]), where $C_n = \binom{2n}{n}/(n+1) \leq 2^{2n}$ (see [3]). It is clear that the circuit S is reachable. Using Remark 3.4, we obtain that in the integer model of computation, for the considered scheme of the circuit, the algorithm $\mathcal{A}_2$ has the time complexity $O(n^4 \cdot \log n \cdot \log\log n)$, which is polynomial depending on the length of the problem representation.

We can use the algorithm $\mathcal{A}_4$ to find the set of Pareto optimal points for the problem of bi-criteria optimization of binary search trees relative to Ψ and Φ.

By Proposition 4.4, $M_{\Psi,\Phi}(S) \leq \nu n + 1$, where

$$M_{\Psi,\Phi}(S) = \max\{|Par(t_{\Psi,\Phi}(S, v))| : v \in V(S)\} .$$

Denote $M = M_{\Psi,\Phi}(S)$.

To evaluate the number of operations of the algorithm $\mathscr{A}_4$, we use Theorem 4.1. To treat the input nodes of the scheme of S, we do not need any operations.

The scheme of the circuit S contains at most $(n+1)^3$ functional nodes. Each such node v has two entering edges. It requires at most eight operations to compute the functions ψ_v and φ_v. Also, we need to compute the prefix sum sequences P_1 and P_2, which requires at most $2(n+1)$ operations of addition. Therefore, the algorithm needs at most $8(n+1)^3M^2 + 2(n+1) \leq 10(n+1)^3M^2$ operations from B and at most $8(n+1)^3M^2\log_2 M$ comparisons to process all functional nodes. In total, we need at most $10(n+1)^3M^2 + 8(n+1)^3M^2\log_2 M \leq 18(n+1)^3M^2\log_2(2nM)$ operations from the set B^+.

For each of at most $(n+1)^2$ unifying nodes with at most n entering edges, we make at most $4nM\log_2(nM)$ comparisons. Then, to process all unifying nodes, we need at most $4(n+1)^3M\log_2(nM) \leq 4(n+1)^3M^2\log_2(2nM)$ comparisons.

Hence, the algorithm $\mathscr{A}_4$ makes at most $22(n+1)^3M^2\log_2(2nM)$ operations from the set B^+.

Since $M \leq \nu n + 1$, the algorithm $\mathscr{A}_4$ makes at most $44(n+1)^5\nu^2\log_2(2(n+1)\nu)$ operations from the set B^+, i.e., pseudo-polynomial number of operations (polynomial depending on the dimension n and the numeric parameter ν).

In the software model of computation, for the considered scheme of the circuit and pair of cost functions, the algorithm $\mathscr{A}_4$ has the time complexity

$$O(n^5\nu^2\log(n\nu)) ,$$

which is pseudo-polynomial (polynomial depending on the dimension n and the numeric parameter ν).

We have only one integer cost function Ψ. Therefore, we cannot consider the integer model of computation in the case of the algorithm $\mathscr{A}_4$.

8.3 Example

Let us consider an example of three keys k_1, k_2, k_3 with associated weights $w_1 = 1, w_2 = 2, w_3 = 5$ and four dummy keys d_0, d_1, d_2, d_3. The keys and dummy keys have the probabilities $p_1 = 0.2, p_2 = 0.2, p_3 = 0.2$ and $q_0 = 0.1, q_1 = 0.1, q_2 = 0.1, q_3 = 0.1$. The set of all binary search trees for the considered example is described by the circuit S_{BST} depicted in Fig. 8.3.

First, we apply to the circuit S_{BST} the optimization procedure (algorithm $\mathscr{A}_1$) relative to the average weighted depth Φ. As a result, we construct the circuit S_{BST}^{Φ}, which can be obtained from S_{BST} by the removal of edges labeled with 1. The

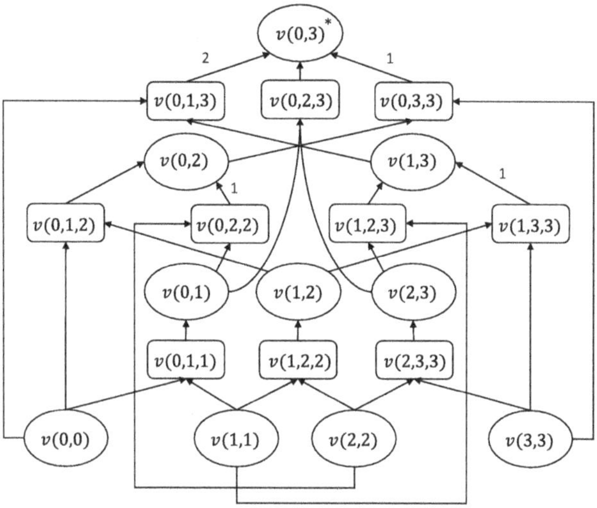

Fig. 8.3 Circuit S_{BST}

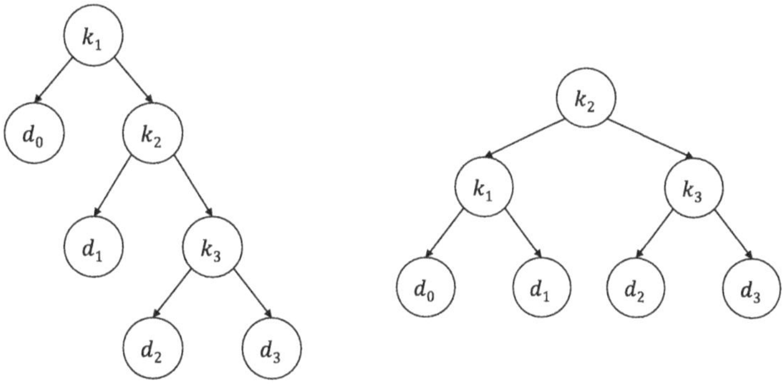

Fig. 8.4 Binary search trees corresponding to node $v(0, 3)$ of the circuit S^{Φ}_{BST}

binary search trees depicted in Fig. 8.4 correspond to the output node $v(0, 3)$ of the circuit S^{Φ}_{BST}. These are the only two optimal binary search trees relative to Φ for the considered example.

Next, we apply to the circuit S^{Φ}_{BST} the procedure of optimization relative to the weighted depth Ψ. As a result, we construct the circuit $S^{\Phi,\Psi}_{BST}$, which can be obtained from S_{BST} by the removal of edges labeled with 1 and 2. The circuit $S^{\Phi,\Psi}_{BST}$ describes the only one binary search tree (the right one in Fig. 8.4), which is optimal relative to Ψ among all trees optimal relative to Φ.

We also apply the bi-criteria optimization procedure (algorithm $\mathscr{A}_4$) relative to Ψ and Φ. As a result, we obtain the set of Pareto optimal points for each node of the circuit S_{BST}—see Fig. 8.5.

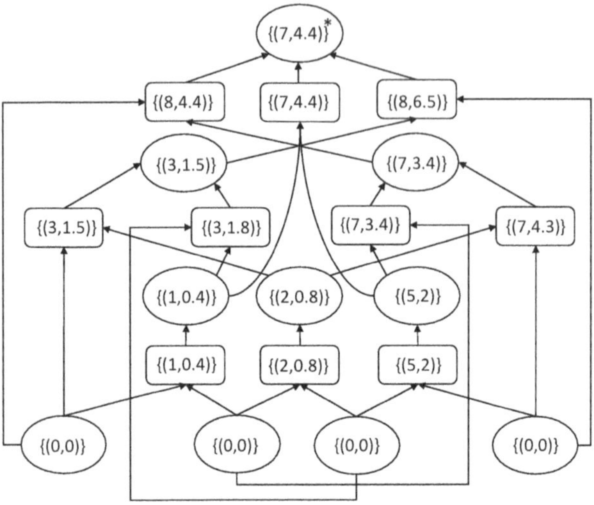

Fig. 8.5 Circuit S_{BST} and corresponding sets of Pareto optimal points

8.4 Experiments

We experimentally tested different optimization scenarios on binary search trees. Each experiment was repeated 100 times.

The number of keys n was fixed for a random trial. The probabilities $p_1, p_2, \ldots, p_n$ and $q_0, q_1, \ldots, q_n$ were obtained by generating $2n + 1$ uniformly distributed integers from 0 to 100. Then each of the random numbers was divided by the sum of all $2n + 1$ integers. Also, each key had randomly assigned a weight from 1 to some value.

Table 8.1 shows, for $n = 10, 20, \ldots, 100$, the number of binary search trees described by the circuit before optimization, after optimization relative to Ψ, relative to Φ, relative to Ψ, Φ, and relative to Φ, Ψ. In the subtable A, we consider weights from 1 to 2, and in the subtable B, we consider weights from 1 to 100. Increasing the maximum value of key weights from 2 to 100 resulted in a decrease in the number of optimal binary search trees relative to Ψ. The number of optimal binary search trees relative to Φ was one for each of the considered trials. It is expectable since the values of Φ are floating-point numbers.

Table 8.2 shows the average value of the function Ψ after optimization relative to Ψ, and relative to Φ, Ψ, and the average value of the function Φ after optimization relative to Φ, and relative to Ψ, Φ. The two-stage optimization resulted in the increase of average values in comparison with the one-stage optimization.

We also consider a bi-criteria optimization of binary search trees. Table 8.3 shows the minimum, maximum, and the average number of Pareto optimal points for the bi-criteria optimization problem relative to Φ and Ψ for $n = 10, 20, \ldots, 100$ keys and maximum weights $10, 30, \ldots, 110$.

Table 8.1 Average number of binary search trees described by the circuit before optimization, after optimization relative to Ψ, relative to Φ, relative to Ψ, Φ, and relative to Φ, Ψ among 100 trials with weights from 1 to 2 (A) and from 1 to 100 (B)

A

n	$\emptyset$	Ψ	Φ	Ψ, Φ	Φ, Ψ
10	1.68×10^{04}	26.08	1	1	1
20	6.56×10^{09}	1.35×10^{03}	1	1	1
30	3.81×10^{15}	7.26×10^{04}	1	1	1
40	2.62×10^{21}	4.96×10^{06}	1	1	1
50	1.98×10^{27}	3.62×10^{08}	1	1	1
60	1.58×10^{33}	3.69×10^{10}	1	1	1
70	1.32×10^{39}	2.30×10^{12}	1	1	1
80	1.14×10^{45}	6.85×10^{13}	1	1	1
90	1.00×10^{51}	5.90×10^{15}	1	1	1
100	8.97×10^{56}	2.96×10^{17}	1	1	1

B

n	$\emptyset$	Ψ	Φ	Ψ, Φ	Φ, Ψ
10	1.68×10^{04}	4.59	1	1	1
20	6.56×10^{09}	22.72	1	1	1
30	3.81×10^{15}	77.14	1	1	1
40	2.62×10^{21}	2.67×10^{02}	1	1	1
50	1.98×10^{27}	1.35×10^{03}	1	1	1
60	1.58×10^{33}	3.29×10^{03}	1	1	1
70	1.32×10^{39}	4.09×10^{04}	1	1	1
80	1.14×10^{45}	2.08×10^{05}	1	1	1
90	1.00×10^{51}	5.76×10^{05}	1	1	1
100	8.97×10^{56}	1.88×10^{06}	1	1	1

Table 8.2 Average value of Ψ after optimization relative to Ψ and Φ, Ψ, and average value of Φ after optimization relative to Φ and Ψ, Φ among 100 trials with weights from 1 to 100

n	Ψ	Φ, Ψ	Φ	Ψ, Φ
10	2.02×10^{02}	2.36×10^{02}	1.14×10^{02}	1.35×10^{02}
20	2.36×10^{02}	2.86×10^{02}	1.28×10^{02}	1.58×10^{02}
30	2.51×10^{02}	3.10×10^{02}	1.35×10^{02}	1.66×10^{02}
40	2.61×10^{02}	3.30×10^{02}	1.40×10^{02}	1.73×10^{02}
50	2.69×10^{02}	3.38×10^{02}	1.43×10^{02}	1.75×10^{02}
60	2.74×10^{02}	3.48×10^{02}	1.44×10^{02}	1.80×10^{02}
70	2.78×10^{02}	3.56×10^{02}	1.46×10^{02}	1.82×10^{02}
80	2.81×10^{02}	3.58×10^{02}	1.48×10^{02}	1.85×10^{02}
90	2.82×10^{02}	3.62×10^{02}	1.48×10^{02}	1.87×10^{02}
100	2.85×10^{02}	3.66×10^{02}	1.48×10^{02}	1.88×10^{02}

Table 8.3 Number of Pareto optimal points (in format ${}_{min}average_{max}$ among 100 trials) for bi-criteria optimization problem relative to Φ and Ψ

n	Maximum weight					
	10	30	50	70	90	110
10	${}_{1}2.71_{6}$	${}_{1}3.21_{11}$	${}_{1}3.26_{9}$	${}_{1}3.14_{9}$	${}_{1}3.33_{7}$	${}_{1}3.84_{15}$
20	${}_{1}4.28_{13}$	${}_{1}4.78_{11}$	${}_{2}5.59_{11}$	${}_{1}5.62_{14}$	${}_{2}5.87_{16}$	${}_{1}5.92_{20}$
30	${}_{1}5.04_{9}$	${}_{2}6.50_{15}$	${}_{1}7.43_{19}$	${}_{1}6.86_{13}$	${}_{1}8.06_{27}$	${}_{1}7.75_{22}$
40	${}_{2}5.71_{11}$	${}_{3}7.93_{17}$	${}_{2}8.66_{24}$	${}_{1}9.03_{20}$	${}_{2}9.01_{22}$	${}_{3}10.76_{26}$
50	${}_{3}5.94_{14}$	${}_{2}8.94_{16}$	${}_{3}10.26_{19}$	${}_{2}10.19_{27}$	${}_{3}12.04_{28}$	${}_{2}10.65_{24}$
60	${}_{3}6.37_{11}$	${}_{2}9.77_{16}$	${}_{3}11.50_{25}$	${}_{4}12.14_{26}$	${}_{3}11.41_{23}$	${}_{4}12.94_{27}$
70	${}_{2}6.91_{13}$	${}_{4}10.10_{24}$	${}_{6}12.74_{28}$	${}_{5}13.68_{25}$	${}_{4}13.17_{28}$	${}_{4}14.63_{28}$
80	${}_{3}7.11_{12}$	${}_{4}11.64_{20}$	${}_{4}13.55_{28}$	${}_{6}13.90_{35}$	${}_{5}15.01_{38}$	${}_{5}15.26_{33}$
90	${}_{3}7.29_{12}$	${}_{5}11.95_{25}$	${}_{5}13.29_{24}$	${}_{4}15.27_{32}$	${}_{7}15.88_{34}$	${}_{4}16.20_{35}$
100	${}_{3}7.75_{16}$	${}_{5}12.34_{22}$	${}_{4}14.31_{29}$	${}_{4}14.98_{26}$	${}_{4}16.10_{27}$	${}_{4}17.60_{40}$

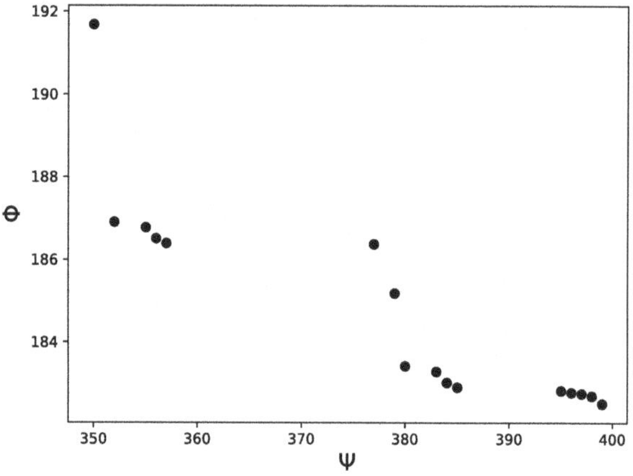

Fig. 8.6 Set of Pareto optimal points for a bi-criteria optimization of binary search trees with 100 keys and weights from 1 to 100

With the increase in the number of keys and maximum weight, the average number of Pareto optimal points increased in most of the cases. An example of the set of Pareto optimal points for an optimization problem with 100 keys and weights from 1 to 100 is depicted in Fig. 8.6.

The number of totally optimal binary search trees relative to Φ and Ψ was zero for most of the evaluated problems. Only for the problems with ten keys, there were totally optimal binary search trees in each tested weighting scenario (see Table 8.4).

Table 8.4 Number of cases with totally optimal binary search trees relative to Φ and Ψ among 100 trials

n	Maximum weights					
	10	30	50	70	90	110
10	12	13	11	14	9	8
20	1	1	0	2	0	2
30	1	0	1	1	1	1
40	0	0	0	1	0	0
50	0	0	0	0	0	0
60	0	0	0	0	0	0
70	0	0	0	0	0	0
80	0	0	0	0	0	0
90	0	0	0	0	0	0
100	0	0	0	0	0	0

References

1. Alnafie, M., Chikalov, I., Hussain, S., Moshkov, M.: Sequential optimization of binary search trees for multiple cost functions. In: Potanin, A., Viglas, T. (eds.) Seventeenth Computing: The Australasian Theory Symposium, CATS 2011, Perth, Australia, January 17–20, 2011, CRPIT, vol. 119, pp. 41–44. Australian Computer Society (2011)
2. Cormen, T.H., Stein, C., Rivest, R.L., Leiserson, C.E.: Introduction to Algorithms, 2nd edn. The MIT Press and McGraw-Hill Book Company (2001)
3. Dutton, R.D., Brigham, R.C.: Computationally efficient bounds for the Catalan numbers. Eur. J. Comb. **7**(3), 211–213 (1986)
4. Knuth, D.E.: Optimum binary search trees. Acta Inform. **1**(1), 14–25 (1971)
5. Mankowski, M., Moshkov, M.: Dynamic programming bi-criteria combinatorial optimization. Discret. Appl. Math. **284**, 513–533 (2020)
6. Stanley, R.P.: Enumerative Combinatorics. Cambridge Studies in Advanced Mathematics, vol. 2. Cambridge University Press (1999)
7. Szwarcfiter, J.L., Navarro, G., Baeza-Yates, R., de S. Oliveira, J., Cunto, W., Ziviani, N.: Optimal binary search trees with costs depending on the access paths. Theor. Comput. Sci. **290**(3), 1799–1814 (2003)

Chapter 9
Convex Polygon Triangulation

A convex polygon triangulation is a decomposition of the polygon into a set of triangles with pairwise non-intersecting interiors whose union is this polygon. We consider only triangles in which vertices are vertices of the polygon.

A dynamic programming algorithm for the optimization of a convex polygon triangulation was introduced independently by Gilbert in 1979 [3] and by Klincsek in 1980 [5]. Both authors considered the minimization of the sum of the lengths of edges. Also, other well-known objectives were studied such as the min-max length triangulation, which minimizes the length of the longest edge of triangles [2], and the max-min area triangulation, which maximizes the minimum area of a triangle in the triangulation [4].

In this chapter, we construct a conventional circuit without repetitions that represents all triangulations of a given polygon, describe three cost functions for this circuit, evaluate the number of operations and the time required by the algorithms, consider an example of a polygon, and discuss the results of experiments with the algorithms $\mathscr{A}_1$, $\mathscr{A}_2$, and $\mathscr{A}_4$ for randomly generated polygons. This chapter is partially based on [6].

9.1 Definition of Circuit and Cost Functions

Let $P(1, n)$ be a convex polygon with the set of $n + 1$ vertices $V = \{v_0, v_1, \ldots, v_n\}$, and v_a, v_b, v_c be pairwise different vertices from V. We denote by $\triangle v_a v_b v_c$ the triangle with vertices v_a, v_b, v_c and edges (v_a, v_b), (v_b, v_c), (v_c, v_a). A triangulation of the polygon $P(1, n)$ is a set of triangles $\triangle v_a v_b v_c$, $v_a, v_b, v_c \in V$, with pairwise non-intersecting interiors whose union is $P(1, n)$.

The number n is the dimension of the problems under consideration related to the optimization of convex polygon triangulations. To describe an instance of a problem,

M. Mankowski and M. Moshkov, *Dynamic Programming Multi-Objective Combinatorial Optimization*, Studies in Systems, Decision and Control 331,
https://doi.org/10.1007/978-3-030-63920-4_9

we need to specify coordinates of $n + 1$ vertices of a convex polygon, i.e., $2(n + 1)$ real numbers. Note that n is at most the number of numeric parameters of the problem.

Let us assume that the coordinates of all vertices are integers. The binary representations of these $2(n + 1)$ integers form the representation of the problem instance (we will omit the word instance). It is clear that n is at most the length of the problem representation.

For any nonempty sets of triangulations C and D, we denote $C \otimes D = \{c \cup d : c \in C, d \in D\}$. For any $i, j, 1 \leq i \leq j \leq n$, we describe the set $T(i, j)$ of all triangulations of the polygon $P(i, j)$, which is the convex hull of vertices $v_{i-1}, v_i, \ldots, v_j$. If $i = j$, then $P(i, i)$ is a 2-vertex polygon, which by definition, has only one triangulation $\emptyset$, which is the empty set, i.e., $T(i, i) = \{\emptyset\}$. Let $i < j$. Then $T(i, j) = \bigcup_{k=i}^{j-1} T(i, k, j)$, where

$$T(i, k, j) = (T(i, k) \otimes T(k + 1, j)) \cup \{\Delta v_{i-1} v_k v_j\} .$$

Let $U = \bigcup_{1 \leq i \leq j \leq n} T(i, j)$. We define a circuit S without repetitions over the set U, which describes the set $T(1, n)$ of all triangulations of the polygon $P(1, n)$.

For, $i, k, j, 1 \leq i \leq k < j \leq n$, we define a concatenation function $\mu_{i,k,j}$ from the set $T(i, k) \times T(k + 1, j)$ to the set $T(i, k, j)$ such that $\mu_{i,k,j}(t_1, t_2) = t_1 \cup t_2 \cup \{\Delta v_{i-1} v_k v_j\}$, where $t_1 \in T(i, k)$ and $t_2 \in T(k + 1, j)$.

The circuit S contains one input node $v(0, 0)$ labeled with $\{\emptyset\}$. For each $i, j, 1 \leq i \leq j \leq n$, the circuit S contains a unifying node $v(i, j)$ labeled with the symbol $\cup$. For each $i, k, j, 1 \leq i \leq k < j \leq n$, the circuit S contains a functional node $v(i, k, j)$ labeled with the function $\mu_{i,k,j}$.

For $i = 1, \ldots, n$, the unifying node $v(i, i)$ has one entering edge leaving the input node $v(0, 0)$. If $i < j$, the unifying node $v(i, j)$ has $j - i$ entering edges leaving the nodes $v(i, k, j)$, $i \leq k \leq j - 1$. The functional node $v(i, k, j)$ has two entering edges labeled with the numbers 1 and 2 (in the example, we omit these numbers), which leave the nodes $v(i, k)$ and $v(k + 1, j)$. The node $v(1, n)$ is the output of S labeled with $*$. One can show that $S(v(i, k, j)) = T(i, k, j)$ for any $i, k, j, 1 \leq i \leq k < j \leq n$, and $S(v(i, j)) = T(i, j)$ for any $i, j, 1 \leq i \leq j \leq n$.

The described circuit will be considered as the preliminary version of the circuit S. The final version of S is obtained from the preliminary version of S by the removal of all nodes such that there is no directed path from the considered node to the output. We also remove all edges that are incident to the removed nodes. Later, S without mentioning the version denotes the final version of the circuit S.

There is one input node, at most n^2 unifying nodes with at most n entering edges each, and at most n^3 functional nodes with two entering edges each in the preliminary version of the circuit S. This circuit contains at most $1 + n^2 + n^3$ nodes and at most $3n^3$ edges.

Let us show that S is a circuit without repetitions. It is clear that the functions $\mu_{i,k,j}$ are injective. Let v_1 and v_2 be different nodes from $V_i(S) \cup V_f(S)$. We now show that $S(v_1) \cap S(v_2) = \emptyset$. Let v_1 be the input node and v_2 be a functional node. Then $S(v_1)$ contains the only empty set, but $S(v_2)$ does not contain any empty set, hence $S(v_1) \cap S(v_2) = \emptyset$. Let v_1 and v_2 be functional nodes, $v_1 = v(i_1, k_1, j_1)$ and

$v_2 = v(i_2, k_2, j_2)$. If $(i_1, j_1) = (i_2, j_2)$ and $k_1 \neq k_2$, then $\Delta v_{i_1-1} v_{k_1} v_{j_1} \in t_1$ for each $t_1 \in S(v_1)$, but $\Delta v_{i_1-1} v_{k_1} v_{j_1} \notin t_2$ for any $t_2 \in S(v_2)$. It is clear that $S(v_1) \cap S(v_2) = \emptyset$. If $(i_1, j_1) \neq (i_2, j_2)$, then $S(v_1)$ and $S(v_2)$ are sets of triangulations of two different polygons. Evidently, $S(v_1) \cap S(v_2) = \emptyset$.

We now define the notion of a cost function $\Psi^{(t)}$, $t = 1, 2, 3$, for the circuit S given by a function $\psi^{(t)}$ and function $\psi_v^{(t)}$ for each $v \in V_f(S)$. Denote

$$B = \{x + y, x - y, x \cdot y, x/y, \sqrt{x}, \max(x, y), |x|\} .$$

Each vertex $v_i \in V$ is a point (a_i, b_i) in the two-dimensional Euclidean space.

The cost function $\Psi^{(1)}$ (length of the longest edge in a triangulation) is defined by functions $\psi^{(1)}(\emptyset) = 0$ and

$$\psi^{(1)}_{v(i,k,j)}(x_1, x_2) = \max\{x_1, x_2, l(v_{i-1}, v_k), l(v_k, v_j), l(v_j, v_{i-1})\} ,$$

$v(i, k, j) \in V_f(S)$, where $l(v_p, v_q)$ is the length of the edge between nodes v_p, v_q. To calculate $l(v_p, v_q)$, we need six operations from B using the formula

$$\sqrt{(a_p - a_q)^2 + (b_p - b_q)^2} .$$

To compute the function $\psi^{(1)}_{v(i,k,j)}$, we need 22 operations from B. The cost function $\Psi^{(1)}$ is increasing.

The cost function $\Psi^{(2)}$ (sum of the lengths of edges of all triangles in a triangulation) is defined by functions $\psi^{(2)}(\emptyset) = 0$ and $\psi^{(2)}_{v(i,k,j)}(x_1, x_2) = x_1 + x_2 + l(v_{i-1}, v_k) + l(v_k, v_j) + l(v_j, v_{i-1})$, $v(i, k, j) \in V_f(S)$. To compute the function $\psi^{(2)}_{v(i,k,j)}$, we need 22 operations from B. The cost function $\Psi^{(2)}$ is strictly increasing.

Let $A_{\Delta v_{i-1} v_k v_j}$ be the area of the triangle $\Delta v_{i-1} v_k v_j$. The cost function $\Psi^{(3)}$ (negation of the minimum area of a triangle in a triangulation) is defined by functions $\psi^{(3)}(\emptyset) = -\infty$ and $\psi^{(3)}_{v(i,k,j)}(x_1, x_2) = \max\{x_1, x_2, -A_{\Delta v_{i-1} v_k v_j}\}$, $v(i, k, j) \in V_f(S)$. To calculate the area of a triangle, we use the formula

$$A_{\Delta v_{i-1} v_k v_j} = \frac{1}{2} |a_{i-1}(b_k - b_j) + a_k(b_j - b_{i-1}) + a_j(b_{i-1} - b_k)| .$$

To compute the function $\psi^{(3)}_{v(i,k,j)}$, we need 13 operations from B. The cost function $\Psi^{(3)}$ is increasing. We have no tools for the maximization of cost functions. So instead of the maximization of the minimum area of a triangle in a triangulation, we minimize the negation of the minimum area of a triangle in a triangulation.

All these cost functions are adapted (see Sect. 2.4). They are described by formulas over the set of elementary operations B.

9.2 Time Complexity of Algorithms

The algorithms $\mathscr{A}_1$, $\mathscr{A}_2$, and $\mathscr{A}_4$ work with the scheme of the circuit S (see Sect. 2.4). The preliminary version of the circuit S is completely described. Therefore, the time complexity of the scheme of the preliminary version of S construction is linear depending on the number of its nodes and edges, i.e., $O(n^3)$. Using breadth-first search from the output node in the scheme of the preliminary version of S with the reverse direction of the edges, we can find in the scheme of the preliminary version of S all nodes such that there is no directed path from the considered node to the output and construct the scheme of S in $O(n^3)$ time. In total, we need $O(n^3)$ time to construct the scheme of S.

Note that the algorithms $\mathscr{A}_1$ and $\mathscr{A}_2$ can also work with the schemes of edge-preserving subcircuits of S obtained as a result of the algorithm $\mathscr{A}_1$ work—see Remark 3.1.

Denote $B^+ = B \cup \{x : y\}$, where $x : y$ is the operation of comparison with values $x < y$, $x = y$, and $x > y$.

Note that $\Psi^{(1)}$, $\Psi^{(2)}$, and $\Psi^{(3)}$ are not integer cost functions. So we will not consider the integer model of computation for the algorithms $\mathscr{A}_1$ and $\mathscr{A}_4$.

We can use the algorithm $\mathscr{A}_1$ for multi-stage optimization of convex polygon triangulations. Using Theorem 3.1, we evaluate the number of operations that the algorithm $\mathscr{A}_1$ makes when works with the scheme of an edge-preserving subcircuit T of the circuit S and a cost function $\Psi^{(t)}$ for S, where $t \in \{1, 2, 3\}$.

The scheme of T contains one input node. We do not need any operations to process this node. The scheme of T contains at most n^2 unifying nodes. Every unifying node has at most n entering edges. The algorithm $\mathscr{A}_1$ makes at most $2n - 1$ comparisons per node and at most $(2n - 1)n^2 \leq 2n^3$ comparisons for all unifying nodes.

The scheme of T contains at most n^3 functional nodes. For each functional node v, it is enough to make at most 22 operations from B to compute $\psi_v^{(t)}$. The number of operations to treat all functional nodes is at most $22n^3$. The total number of operations from the set B^+ required by $\mathscr{A}_1$ is at most $24n^3$, i.e., it is polynomial depending on the problem dimension n.

In the software model of computation, for the considered scheme of the circuit and cost function, the algorithm $\mathscr{A}_1$ has polynomial time complexity $O(n^3)$ depending on the problem dimension n.

To count all possible triangulations described by the circuit T, we apply the algorithm $\mathscr{A}_2$. We use Theorem 3.2 to evaluate the number of operations required by $\mathscr{A}_2$. The scheme of T contains at most n^2 unifying nodes with at most n entering edges each. Therefore, we need at most $(n - 1)n^2 \leq n^3$ additions to treat all unifying nodes. To process one functional node, it is enough to make one multiplication since each functional node has two entering edges. To treat all functional nodes, we need at most n^3 multiplications. The input node does not require any operations. The total number of operations from the set B^+ required by $\mathscr{A}_2$ is at most $2n^3$, i.e., it is polynomial depending on the problem dimension n.

In the software model of computation, for the considered scheme of the circuit, the algorithm $\mathscr{A}_2$ has polynomial time complexity $O(n^3)$ depending on the problem dimension n.

We now evaluate the number $|S(v(1,n))|$, which is equal to the number of all triangulations of the polygon $P(1,n)$. This value is equal to the $(n-1)$th Catalan number C_{n-1} (see [7]), where $C_n = \binom{2n}{n}/(n+1) \leq 2^{2n}$ (see [1]). It is clear that the circuit S is reachable. Using Remark 3.4, we obtain that in the integer model of computation, for the considered scheme of the circuit, the algorithm $\mathscr{A}_2$ has time complexity $O(n^4 \cdot \log n \cdot \log\log n)$, which is polynomial depending on the length of the problem representation when the coordinates of all vertices are integers.

We can use the algorithm $\mathscr{A}_4$ to find the set of Pareto optimal points for the problem of bi-criteria optimization of triangulations relative to cost functions $\Psi^{(i)}$ and $\Psi^{(j)}$, $i \in \{1,3\}$, $j \in \{1,2,3\}$, and $i \neq j$. Denote $M = M_{\Psi^{(i)},\Psi^{(j)}}(S)$, where $M_{\Psi^{(i)},\Psi^{(j)}}(S) = \max\{|Par(t_{\Psi^{(i)},\Psi^{(j)}}(S,v))| : v \in V(S)\}$.

To evaluate the number of operations of the algorithm $\mathscr{A}_4$, we use Theorem 4.1. To treat the input node of the scheme of S, we do not need any operations.

The scheme of the circuit S contains at most n^3 functional nodes. Each such node v has two entering edges. It requires at most 44 operations from B to compute the functions $\psi_v^{(i)}$ and $\psi_v^{(j)}$. Therefore, the algorithm needs at most $44n^3M^2$ operations from B and at most $8n^3M^2\log_2 M$ comparisons to process all functional nodes. In total, we need at most $44n^3M^2 + 8n^3M^2\log_2 M \leq 52n^3M^2\log_2(nM)$ operations from the set B^+.

For each of at most n^2 unifying nodes with at most n entering edges, we make at most $4nM\log_2(nM)$ comparisons. Then, to process all unifying nodes, we need at most $4n^3M\log_2(nM) \leq 4n^3M^2\log_2(nM)$ comparisons.

Hence, the algorithm $\mathscr{A}_4$ makes at most $56n^3M^2\log_2(nM)$ operations from the set B^+.

Let $\Psi^{(i)} = \Psi^{(1)}$. Then $|\{\Psi^{(1)}(\delta) : \delta \in U(S)\}| \leq (n+1)^2$ since the number of different edges in the triangulations of the polygon $P(1,n)$ is at most $(n+1)^2$. By Proposition 4.5, $M_{\Psi^{(i)},\Psi^{(j)}}(S) \leq (n+1)^2$. Since $M \leq (n+1)^2$, the algorithm $\mathscr{A}_4$ makes at most $56n^3(n+1)^4\log_2(n(n+1)^2) \leq 168(n+1)^7\log_2(n+1)$ operations from B^+, i.e., polynomial number of operations depending on the dimension n.

Let $\Psi^{(i)} = \Psi^{(3)}$. Then $|\{\Psi^{(3)}(\delta) : \delta \in U(S)\}| \leq (n+1)^3$ since the number of different triangles in the triangulations of the polygon $P(1,n)$ is at most $(n+1)^3$. By Proposition 4.5, $M_{\Psi^{(i)},\Psi^{(j)}}(S) \leq (n+1)^3$. Since $M \leq (n+1)^3$, the algorithm $\mathscr{A}_4$ makes at most $56n^3(n+1)^6\log_2(n(n+1)^3) \leq 224(n+1)^9\log_2(n+1)$ operations from B^+, i.e., polynomial number of operations depending on the dimension n.

In the software model of computation, for the considered scheme of the circuit and pair of cost functions, the algorithm $\mathscr{A}_4$ has the time complexity $O(n^7\log n)$ if $\Psi^{(i)} = \Psi^{(1)}$ and $O(n^9\log n)$ if $\Psi^{(i)} = \Psi^{(3)}$, which is polynomial depending on the dimension n.

9.3 Example

We consider a polygon $P(1,4)$ with five vertices $v_0 = (0,0)$, $v_1 = (1,0)$, $v_2 = (2,1)$, $v_3 = (1,2)$, and $v_4 = (0,2)$—see Fig. 9.1.

The circuit S_{PT} describes the set $T(1,4)$ of all triangulations of the polygon $P(1,4)$ (see Fig. 9.2).

We apply to the circuit S_{PT} the procedure of optimization relative to $\Psi^{(1)}$. As a result, we obtain the circuit $S_{PT}^{\Psi^{(1)}} = S_{PT}$. Both circuits describe all five triangulations of the polygon $P(1,4)$—see Fig. 9.3.

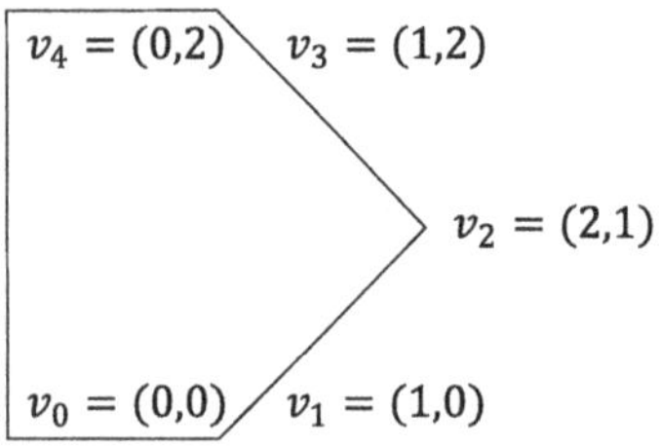

Fig. 9.1 Polygon $P(1,4)$ with five vertices

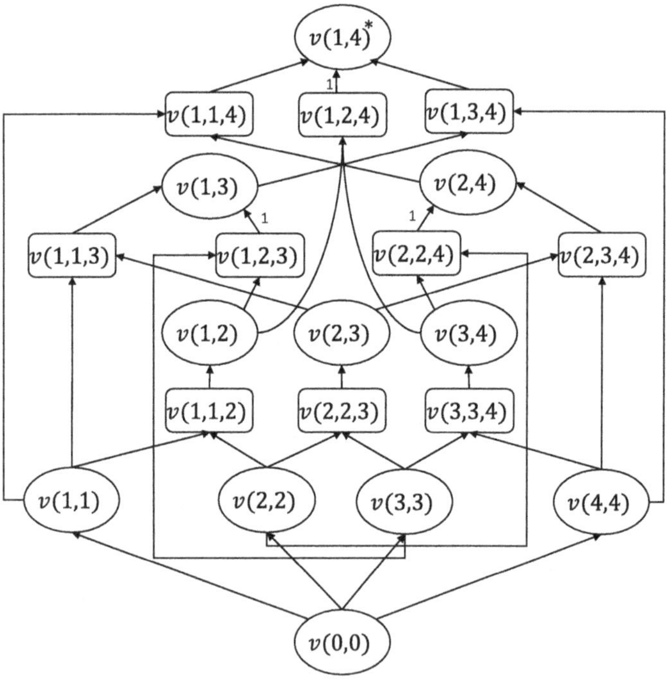

Fig. 9.2 Circuit S_{PT} for the polygon $P(1,4)$

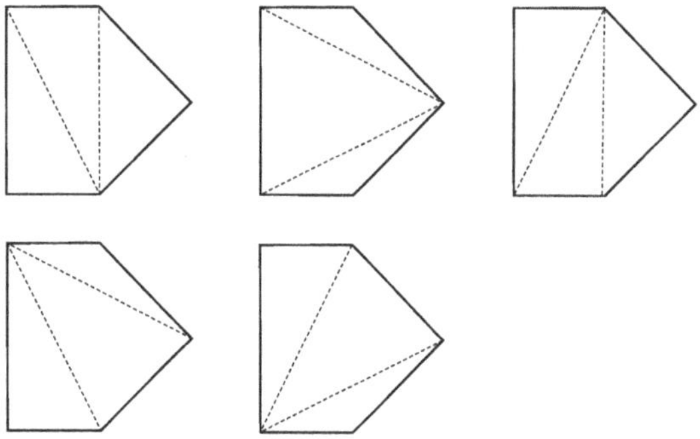

Fig. 9.3 All triangulations of the polygon $P(1,4)$

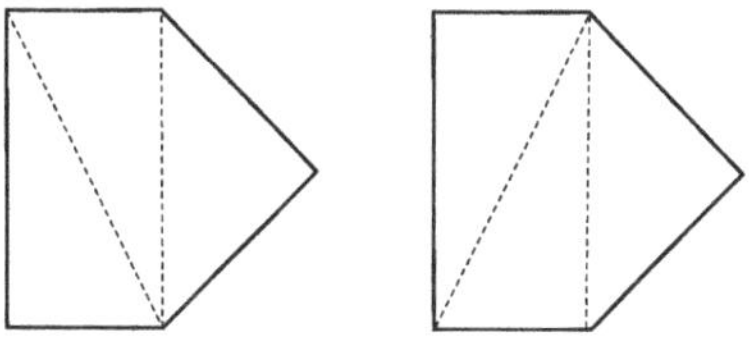

Fig. 9.4 Triangulations represented by the circuit $S_{PT}^{\Psi^{(1)},\Psi^{(2)}}$

Next, we apply to the circuit $S_{PT}^{\Psi^{(1)}}$ the procedure of optimization relative to $\Psi^{(2)}$. As a result, we have the circuit $S_{PT}^{\Psi^{(1)},\Psi^{(2)}}$, which can be obtained from the circuit S_{PT} (see Fig. 9.2) by the removal of edges labeled with the number 1. The circuit $S_{PT}^{\Psi^{(1)},\Psi^{(2)}}$ represents exactly two triangulations depicted in Fig. 9.4.

We also apply the procedure of bi-criteria optimization relative to $\Psi^{(1)}$ and $\Psi^{(2)}$ to the circuit S_{PT}. For each node of S_{PT}, we constructed the set of Pareto optimal points (see Fig. 9.5).

9.4 Experiments

We experimentally tested various optimization scenarios for convex polygon triangulations. Each experiment was repeated 10 or 100 times.

We randomly generated $n+1$ pairwise different points in the circle of radius 1 with the center in the origin and considered these points as vertices of a convex polygon.

Table 9.1 shows for $n = 10, 20, \ldots, 100$, the average number of triangulations described by the circuit before optimization, after optimization relative to $\Psi^{(1)}$, relative to $\Psi^{(2)}$, relative to $\Psi^{(3)}$, relative to $\Psi^{(1)}$, $\Psi^{(3)}$, and relative to $\Psi^{(3)}$, $\Psi^{(1)}$. The number of optimal triangulations relative to $\Psi^{(2)}$ was one for all tested scenarios.

Table 9.2 shows the average value of $\Psi^{(2)}$ after optimization relative to $\Psi^{(2)}$ and $\Psi^{(3)}$, $\Psi^{(2)}$, and average value of $\Psi^{(3)}$ after optimization relative to $\Psi^{(3)}$ and $\Psi^{(2)}$, $\Psi^{(3)}$. We see a significant difference in the results of optimization relative to $\Psi^{(3)}$ and $\Psi^{(2)}$, $\Psi^{(3)}$.

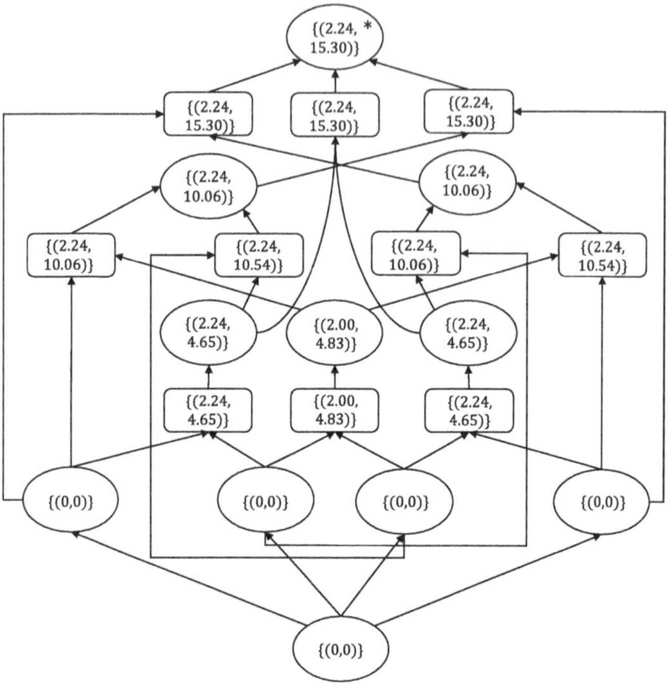

Fig. 9.5 Sets of Pareto optimal points for all nodes of the circuit S_{PT}

Table 9.1 Average number of polygon triangulations described by the circuit before optimization, after optimization relative to $\Psi^{(1)}$, relative to $\Psi^{(2)}$, relative to $\Psi^{(3)}$, relative to $\Psi^{(1)}$, $\Psi^{(3)}$, and relative to $\Psi^{(3)}$, $\Psi^{(1)}$ among 100 trials

n	$\emptyset$	$\Psi^{(1)}$	$\Psi^{(2)}$	$\Psi^{(3)}$	$\Psi^{(1)}, \Psi^{(3)}$	$\Psi^{(3)}, \Psi^{(1)}$
10	4.86×10^{03}	2.19×10^{03}	1	24.43	12.54	22.11
20	1.77×10^{09}	2.64×10^{08}	1	7.37×10^{04}	7.66×10^{03}	1.77×10^{04}
30	1.00×10^{15}	7.08×10^{13}	1	1.07×10^{08}	4.87×10^{06}	5.28×10^{07}
40	6.80×10^{20}	2.88×10^{19}	1	6.68×10^{11}	5.27×10^{10}	1.61×10^{11}
50	5.10×10^{26}	7.89×10^{24}	1	2.05×10^{15}	7.01×10^{15}	5.97×10^{14}
60	4.06×10^{32}	3.09×10^{30}	1	1.61×10^{20}	1.11×10^{21}	6.40×10^{19}
70	3.37×10^{38}	1.20×10^{36}	1	5.40×10^{24}	8.30×10^{25}	1.89×10^{24}
80	2.89×10^{44}	4.66×10^{41}	1	6.89×10^{28}	3.45×10^{30}	1.71×10^{28}
90	2.54×10^{50}	1.56×10^{47}	1	1.99×10^{33}	1.25×10^{35}	2.56×10^{32}
100	2.28×10^{56}	4.18×10^{52}	1	3.23×10^{37}	1.01×10^{39}	5.44×10^{36}

Table 9.2 Average value of $\Psi^{(2)}$ after optimization relative to $\Psi^{(2)}$ and $\Psi^{(3)}$, $\Psi^{(2)}$, and average value of $\Psi^{(3)}$ after optimization relative to $\Psi^{(3)}$ and $\Psi^{(2)}$, $\Psi^{(3)}$ among 100 trials

n	$\Psi^{(2)}$	$\Psi^{(3)}, \Psi^{(2)}$	$\Psi^{(3)}$	$\Psi^{(2)}, \Psi^{(3)}$
10	2.40×10^{02}	3.93×10^{02}	-19.90	-0.13
20	4.52×10^{02}	7.91×10^{02}	-14.02	-0.03
30	6.82×10^{02}	1.14×10^{03}	-11.96	-5.39×10^{-03}
40	9.20×10^{02}	1.48×10^{03}	-6.83	-1.73×10^{-03}
50	1.14×10^{03}	1.81×10^{03}	-5.71	-1.44×10^{-03}
60	1.36×10^{03}	2.16×10^{03}	-4.68	-6.45×10^{-04}
70	1.60×10^{03}	2.48×10^{03}	-4.68	-3.62×10^{-04}
80	1.82×10^{03}	2.79×10^{03}	-3.92	-2.84×10^{-04}
90	2.05×10^{03}	3.11×10^{03}	-3.53	-2.46×10^{-04}
100	2.28×10^{03}	3.45×10^{03}	-3.57	-2.19×10^{-04}

Table 9.3 Number of Pareto optimal points (in format ${}_{\min}average_{\max}$ among 10 trials) for bi-criteria optimization problem relative to $\Psi^{(2)}$, $\Psi^{(3)}$ and $\Psi^{(1)}$, $\Psi^{(2)}$

n	$\Psi^{(2)}, \Psi^{(3)}$	$\Psi^{(1)}, \Psi^{(2)}$
10	${}_{5}11.90_{19}$	${}_{1}1.00_{1}$
15	${}_{13}34.00_{53}$	${}_{1}1.00_{1}$
20	${}_{7}29.30_{49}$	${}_{1}1.10_{2}$
25	${}_{21}48.80_{71}$	${}_{1}1.30_{3}$
30	${}_{45}76.20_{117}$	${}_{1}1.00_{1}$
35	${}_{25}57.50_{80}$	${}_{1}1.00_{1}$
40	${}_{36}88.80_{140}$	${}_{1}1.00_{1}$
45	${}_{30}74.80_{136}$	${}_{1}1.00_{1}$
50	${}_{36}88.10_{146}$	${}_{1}1.00_{1}$
55	${}_{32}129.50_{183}$	${}_{1}1.00_{1}$
60	${}_{22}80.20_{142}$	${}_{1}1.00_{1}$
65	${}_{20}132.00_{200}$	${}_{1}1.00_{1}$
70	${}_{97}147.80_{203}$	${}_{1}1.00_{1}$

We also consider bi-criteria optimization of convex polygon triangulations. Table 9.3 shows the minimum, maximum, and the average number of Pareto optimal points for bi-criteria optimization problem relative to $\Psi^{(2)}$, $\Psi^{(3)}$ and $\Psi^{(1)}$, $\Psi^{(2)}$ for $n = 10, 15, \ldots, 70$. In all cases, the number of Pareto optimal points was at least five for bi-criteria optimization relative to $\Psi^{(2)}$ and $\Psi^{(3)}$. For bi-criteria optimization relative to $\Psi^{(1)}$ and $\Psi^{(2)}$, the number of Pareto optimal points was one for most of the cases. There were no totally optimal triangulations relative to $\Psi^{(2)}$ and $\Psi^{(3)}$, while

Table 9.4 Number of cases with totally optimal triangulations relative to $\Psi^{(2)}$, $\Psi^{(3)}$ and $\Psi^{(1)}$, $\Psi^{(2)}$ among 10 trials

n	$\Psi^{(2)}, \Psi^{(3)}$	$\Psi^{(1)}, \Psi^{(2)}$
10	0	10
15	0	10
20	0	9
25	0	8
30	0	10
35	0	10
40	0	10
45	0	10
50	0	10
55	0	10
60	0	10
65	0	10
70	0	10

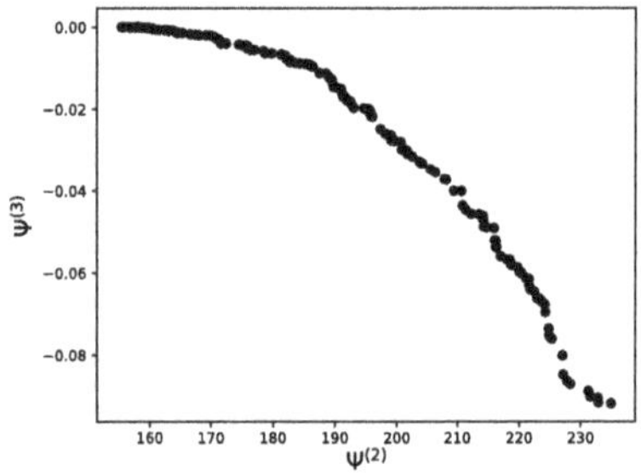

Fig. 9.6 Set of Pareto optimal points for a bi-criteria optimization of convex polygon triangulations ($n = 70$) relative to $\Psi^{(2)}$ and $\Psi^{(3)}$

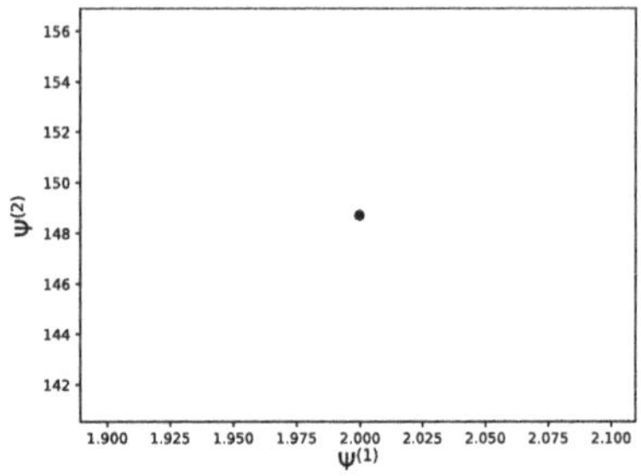

Fig. 9.7 Set of Pareto optimal points for a bi-criteria optimization of convex polygon triangulations ($n = 70$) relative to $\Psi^{(1)}$ and $\Psi^{(2)}$

for most of the cases there were totally optimal triangulations relative to $\Psi^{(1)}$ and $\Psi^{(2)}$ (Table 9.4). The sets of Pareto optimal points for optimization problems with $n = 70$ relative to $\Psi^{(2)}$ and $\Psi^{(3)}$ and relative to $\Psi^{(1)}$ and $\Psi^{(2)}$ are depicted in Fig. 9.6 and in Fig. 9.7.

References

1. Dutton, R.D., Brigham, R.C.: Computationally efficient bounds for the Catalan numbers. Eur. J. Comb. **7**(3), 211–213 (1986)
2. Edelsbrunner, H., Tan, T.S.: A quadratic time algorithm for the minimax length triangulation. SIAM J. Comput. **22**(3), 527–551 (1993)
3. Gilbert, P.D.: New results on planar triangulations. Master's thesis, University of Illinois at Urbana-Champaign (1979)
4. Keil, J.M., Vassilev, T.S.: Algorithms for optimal area triangulations of a convex polygon. Comput. Geom. **35**(3), 173–187 (2006)
5. Klincsek, G.T.: Minimal triangulations of polygonal domains. Ann. Discrete Math. **9**, 121–123 (1980)
6. Mankowski, M., Moshkov, M.: Extensions of dynamic programming for multi-stage combinatorial optimization. Theor. Comput. Sci. **844**, 106–132 (2020)
7. Stanley, R.P.: Enumerative Combinatorics. Cambridge Studies in Advanced Mathematics, vol. 2. Cambridge University Press (1999)

Chapter 10
Line Breaking

Line breaking, word wrap, or text justification problem often relates to the task of text justification into a page with a given width. Despite a type of text alignment (left, right, justified), the text needs to be broken into lines such that any line does not exceed a page width. We refer to this problem as the partitioning of a word sequence into lines with length limit, where word splitting using hyphens is not permitted. This is similar to the process of text left-aligning with ragged right margins. The partitioning of the word sequence into lines is an optimization problem, where one of the objectives is to minimize the sum of badnesses of all lines. The line badness is either square or cube of the length of the remaining empty space at the end of the line. The partitions might also be optimized relative to the number of lines in a partition or to the total penalty for words that should not end the line. The similar version of the line breaking problem with a dynamic programming algorithm for badness minimization was considered in [1]. The extended version of this problem including the dynamic programming algorithm for optimal line breaking in TeX was introduced by Knuth and Plass [2].

In this chapter, we construct a conventional circuit without repetitions that represents all partitions for a given word sequence, describe four cost functions for this circuit, evaluate the number of operations and the time required by the algorithms, consider an example of a word sequence, and discuss the results of experiments with the algorithms $\mathscr{A}_1$, $\mathscr{A}_2$, and $\mathscr{A}_4$ for randomly generated word sequences. This chapter is partially based on [3].

M. Mankowski and M. Moshkov, *Dynamic Programming Multi-Objective Combinatorial Optimization*, Studies in Systems, Decision and Control 331,
https://doi.org/10.1007/978-3-030-63920-4_10

10.1 Definition of Circuit and Cost Functions

Let us have some alphabet A and nonempty words $w_1, w_2, \ldots, w_n$ over $A, n \geq 1$. We consider partitions of this sequence of words into nonempty, non-overlapping, and contiguous lines: $(w_1, \ldots, w_k) \cdots (w_m, \ldots, w_{m+t}) \cdots (w_p, \ldots, w_n)$. We denote by $|w_i|$ the length of the word w_i, $i = 1, \ldots, n$, and assume that these lengths are given. The length $l(w_m, \ldots, w_{m+t})$ of the line $(w_m, \ldots, w_{m+t})$ is equal to $|w_m| + \cdots + |w_{m+t}| + t$, where t stands for the number of gaps between words. Let L be a natural number. We assume that $|w_i| \leq L$ for $i = 1, \ldots, n$ and the length of each line in a partition is at most L.

The number n is the dimension of the problems under consideration related to the optimization of partitions. To describe an instance of a problem, we need to specify the lengths of n words, the maximum possible length of lines L and, probably, 0 or 1 penalty for each word for the case, when this word is at the end of a line, i.e., $n + 1$ natural numbers and, probably, n numbers from the set $\{0, 1\}$. Note that n is at most the number of numeric parameters of the problem.

The binary representations of n lengths of words and the threshold L and, probably, n penalties form the representation of the problem instance (we will omit the word instance). It is clear that n and $\log_2 L$ are at most the length of the problem representation.

We denote by $\alpha \otimes \beta$ the concatenation $\alpha\beta$ of the sequences of lines α and β. Let λ be the empty line. We have $\lambda \otimes \alpha = \alpha$ for any sequence α. For a set C of sequences of lines, we denote $C \otimes \alpha = \{\beta \otimes \alpha : \beta \in C\}$.

For $i = 1, \ldots, n$, we denote by $T(i)$ the set of all partitions of the sequence of words $w_1, \ldots, w_i$. Set $T(0) = \{\lambda\}$. One can show that, for $i = 1, \ldots, n$,

$$T(i) = \bigcup_{\substack{j=0,\ldots,i-1 \\ l(w_{j+1},\ldots,w_i) \leq L}} T(j) \otimes (w_{j+1}, \ldots, w_i) .$$

Set $U = \bigcup_{0 \leq i \leq n} T(i)$. We now describe a circuit S without repetitions over the set U. We define a concatenation function $\mu_{j,i}$, $0 \leq j < i \leq n$, from $T(j)$ to $T(i)$ as follows: $\mu_{j,i}(x) = x \otimes (w_{j+1}, \ldots, w_i)$ for any $x \in T(j)$.

The circuit S contains one input node $v(0)$ labeled with $\{\lambda\}$. For any i, $1 \leq i \leq n$, the circuit S contains a unifying node $v(i)$ labeled with $\cup$. For any $i, j, 0 \leq j < i \leq n$, such that $l(w_{j+1}, \ldots, w_i) \leq L$ the circuit S contains a functional node $v_j(i)$ labeled with the function $\mu_{j,i}$.

The unifying node $v(i)$, $i = 1, \ldots, n$, has at most i entering edges, which leave the functional nodes $v_j(i)$ such that $0 \leq j \leq i - 1$ and $l(w_{j+1}, \ldots, w_i) \leq L$. Each functional node $v_j(i)$ has one entering edge, which leaves $v(j)$. We will not label this edge with the number 1.

One can show that $S(v(i)) = T(i)$ for $i = 1, \ldots, n$ and

$$S(v_j(i)) = T(j) \otimes (w_{j+1}, \ldots, w_i)$$

for any functional node $v_j(i)$ of S. The node $v(n)$ is the output of S labeled with $*$.

The described circuit will be considered as the preliminary version of the circuit S. The final version of S is obtained from the preliminary version of S by the removal of all nodes such that there is no directed path from the considered node to the output. We also remove all edges that are incident to the removed nodes. Later, S without mentioning the version denotes the final version of the circuit S.

The preliminary version of the circuit S contains one input node, n unifying nodes with at most n entering edges each, and at most n^2 functional nodes with one entering edge each. Hence, it has at most $n^2 + n + 1$ nodes and at most $2n^2$ edges.

We now prove that S is a circuit without repetitions. It is clear that the concatenation functions $\mu_{j,i}$ are injective. Let v_1 and v_2 be different nodes from $V_i(S) \cup V_f(S)$. We now show that $S(v_1) \cap S(v_2) = \emptyset$. Let v_1 be the input node and v_2 be a functional node. It is clear that sets $S(v_1)$ and $S(v_2)$ do not contain common elements since $S(v_1)$ contains the only empty line but $S(v_2)$ does not contain any empty line. Let v_1 and v_2 be functional nodes and $v_1 = v_{j_1}(i_1)$, $v_2 = v_{j_2}(i_2)$. If $i_1 \neq i_2$, then $S(v_{j_1}(i_1))$ contains only partitions of the sequence of words $w_1, \ldots, w_{i_1}$ and $S(v_{j_2}(i_2))$ contains only partitions of the sequence of words $w_1, \ldots, w_{i_2}$. Therefore, $S(v_1) \cap S(v_2) = \emptyset$. Let $i_1 = i_2$ and $j_1 \neq j_2$. Then the last line $(w_{j_1+1}, \ldots, w_{i_1})$ in partitions from $S(v_{j_1}(i_1))$ has a different number of words in comparison to the last line $(w_{j_2+1}, \ldots, w_{i_2})$ in partitions from $S(v_{j_2}(i_2))$. Thus, $S(v_1) \cap S(v_2) = \emptyset$.

We now define a cost function $\Psi^{(t)}$, $t = 1, 2, 3, 4$, for the circuit S given by a function $\psi^{(t)}$ and function $\psi_v^{(t)}$ for each $v \in V_f(S)$. Denote

$$B = \{x + y, x - y, x \cdot y\} .$$

Each line has the length limit L. One notion of optimality of the partition is the minimum badness. For a line $(w_m, \ldots, w_{m+t})$, the badness is defined as $(L - l(w_m, \ldots, w_{m+t}))^p$, where $p \in \{2, 3\}$. The badness of a partition is the sum of badnesses of all lines from this partition. The cost function $\Psi^{(1)}$ is defined by the functions $\psi^{(1)}(\lambda) = 0$ and $\psi^{(1)}_{v_j(i)}(x) = x + (L - l(w_{j+1}, \ldots, w_i))^2$, $v_j(i) \in V_f(S)$. Similarly, $\Psi^{(2)}$ is defined by the functions $\psi^{(2)}(\lambda) = 0$ and $\psi^{(2)}_{v_j(i)}(x) = x + (L - l(w_{j+1}, \ldots, w_i))^3$, $v_j(i) \in V_f(S)$. To compute $\psi^{(t)}_{v_j(i)}$, $t \in \{1, 2\}$, knowing $l(w_{j+1}, \ldots, w_i)$, we need at most four elementary operations from the set B.

To calculate, for any i, j, $0 \leq j < i \leq n$, the value $l(w_{j+1}, \ldots, w_i) = |w_{j+1}| + \cdots + |w_i| + i - j - 1$, we use a prefix sum sequence $P = P(0), P(1), \ldots, P(n)$, where $P(0) = 0$ and, for any a, $1 \leq a \leq n$, $P(a) = P(a-1) + |w_a|$. To compute the sequence P, we need at most n elementary operations from the set B. To derive $l(w_{j+1}, \ldots, w_i)$ from P, we need four elementary operations from the set B: $l(w_{j+1}, \ldots, w_i) = P(i) - P(j) + i - j - 1$.

Another objective of optimal partition is to minimize the number of lines that the sequence of words is broken into. The cost function $\Psi^{(3)}$ is defined by the functions $\psi^{(3)}(\lambda) = 0$ and $\psi^{(3)}_{v_j(i)}(x) = x + 1$, $v_j(i) \in V_f(S)$. To compute $\psi^{(3)}_{v_j(i)}$ we need one elementary operation from the set B.

Some parts of speech should not end the line. Typography depends on the language, but for instance, in English, it might not be recommended to end line with the article (a, an, the). We assign to each w_i, $i = 1, 2, \ldots, n$, the line breaking penalty $p_i \in \{0, 1\}$. The cost function $\Psi^{(4)}$ is defined by the functions $\psi^{(4)}(\lambda) = 0$ and $\psi^{(4)}_{v_j(i)}(x) = x + p_i$, $v_j(i) \in V_f(S)$. To compute $\psi^{(4)}_{v_j(i)}$ we need one elementary operation from the set B.

It is clear that $\Psi^{(t)}$, $t = 1, 2, 3, 4$, is a strictly increasing nonnegative integer cost function for S. The considered cost functions are adapted (see Sect. 2.4). They are described by formulas over the set of elementary operations B.

10.2 Time Complexity of Algorithms

The algorithms $\mathscr{A}_1$, $\mathscr{A}_2$, and $\mathscr{A}_4$ work with the scheme of the circuit S (see Sect. 2.4). To describe the scheme of the preliminary version of the circuit S completely, we need to check for any $i, j, 0 \leq j < i \leq n$, if $l(w_{j+1}, \ldots, w_i) \leq L$.

First, we compute the prefix sum sequence P using at most n operations of addition. Then, for each i, j, $0 \leq j < i \leq n$, we compute the value $l(w_{j+1}, \ldots, w_i)$ and check if $l(w_{j+1}, \ldots, w_i) \leq L$. To do this, we need at most $4n^2$ operations of addition and subtraction, and at most n^2 comparisons. As a result, we obtain values $l(w_{j+1}, \ldots, w_i)$, $0 \leq j < i \leq n$, and completely describe the scheme of the preliminary version of the circuit S. In total, we need at most $n + 5n^2$ operations of addition, subtraction, and comparison. The maximum absolute value of the integers at the inputs of these operations is at most $q = \max(|w_1| + \cdots + |w_n| + n, L)$. The time complexity of the considered operations is $O(n^2)$ in the software model of computation and $O(n^2 \log q)$ in the integer model of computation.

Now the scheme of the preliminary version of the circuit S is completely described. The time complexity of this scheme construction is linear depending on the number of its nodes and edges, i.e., $O(n^2)$. Using breadth-first search from the output node in the scheme of the preliminary version of S with the reverse direction of the edges, we can find in the scheme of the preliminary version of S all nodes such that there is no directed path from the considered node to the output and construct the scheme of S in $O(n^2)$ time.

Note that the algorithms $\mathscr{A}_1$ and $\mathscr{A}_2$ can also work with the schemes of edge-preserving subcircuits of S obtained as a result of the algorithm $\mathscr{A}_1$ work—see Remark 3.1.

Denote $B^+ = B \cup \{x : y\}$, where $x : y$ is the operation of comparison with values $x < y$, $x = y$, and $x > y$.

Denote $\rho_1 = L^2$, $\rho_2 = L^3$, $\rho_3 = 1$, and $\rho_4 = \sum_{i=1}^{n} p_i/n$. One can show that, for $t = 1, 2, 3, 4$, $N_{\Psi^{(t)}}(S) \leq \rho_t n$, where $N_{\Psi^{(t)}}(S) = \max\{|\Psi^{(t)}(\delta)| : \delta \in U(S)\}$ and $U(S) = \bigcup_{v \in V(S)} S(v)$. Note that $N_{\Psi^{(t)}}(S) \leq L^3 n$ for any $t \in \{1, 2, 3, 4\}$.

We can use the algorithm $\mathscr{A}_1$ for multi-stage optimization of partitions. Using Theorem 3.1, we evaluate the number of operations from B^+ made by the algorithm

$\mathscr{A}_1$ working with the scheme of an edge-preserving subcircuit T of the circuit S and a cost function $\Psi^{(t)}$ for S, where $t \in \{1, 2, 3, 4\}$.

The scheme of T contains one input node. We do not need any operations to process this node.

The scheme of T contains at most n unifying nodes. Each unifying node has at most n entering edges. To treat all unifying nodes, the algorithm $\mathscr{A}_1$ makes at most $(2n - 1)n \leq 2n^2$ comparisons.

The scheme of T contains at most n^2 functional nodes. For each functional node v, the algorithm makes at most four operations from B^+ to compute $\psi_v^{(t)}$. For all functional nodes, the number of required operations is at most $4n^2$. The total number of operations from B^+ made by $\mathscr{A}_1$ is at most $6n^2$, i.e., it is polynomial depending on the dimension n.

In the software model of computation, for the considered scheme of the circuit and cost function, the algorithm $\mathscr{A}_1$ has polynomial time complexity $O(n^2)$ depending on the problem dimension n.

Using Remark 3.2 we obtain that, for the algorithm $\mathscr{A}_1$, the number L^3n is an upper bound on the absolute values of the numbers at the inputs of the operations $\psi_v^{(t)}$ related to the functional nodes $v \in V_f(S)$ and at the inputs of comparisons related to the unifying nodes $v \in V_u(S)$. It is easy to check that, for the algorithm $\mathscr{A}_1$, the number L^3n is an upper bound on the absolute values of the numbers at the inputs of operations from B in formulas for the functions $\psi_v^{(t)}$ (note that we do not need to recalculate already known values $l(w_{j+1}, \ldots, w_i)$, and all these values used by the algorithm $\mathscr{A}_1$ are at most L). Therefore, in the integer model of computation, for the considered scheme of the circuit and the cost function $\Psi^{(t)}$, the algorithm $\mathscr{A}_1$ has the time complexity $O(n^2 \log(Ln) \log\log(Ln) \log\log\log(Ln))$, which is polynomial depending on the length of the problem representation (n and $\log_2 L$ are at most the length of the problem representation).

To count all possible partitions described by T, we apply the algorithm $\mathscr{A}_2$. We use Theorem 3.2 to evaluate the number of operations from B^+ made by $\mathscr{A}_2$. The scheme of T contains at most n unifying nodes, each with at most n entering edges. Therefore, we need at most n^2 additions to treat all unifying nodes. For each of at most n^2 functional nodes and the input node, we do not need any operations. The total number of operations from B^+ (additions) is at most n^2, i.e., it is polynomial depending on the dimension n.

In the software model of computation, for the considered scheme of the circuit, the algorithm $\mathscr{A}_2$ has polynomial time complexity $O(n^2)$ depending on the problem dimension n.

We now evaluate the number $|S(v(n))|$, which is equal to the number of partitions of the sequence of words $w_1, \ldots, w_n$. One can show that this value is at most 2^{n-1}: there is a natural one-to-one mapping of partitions (without restrictions on the length of each line in a partition) to tuples from $\{0, 1\}^n$ in which the first digit is equal to 1. For example, the tuple $(1, 1, 0, 0, 1)$ corresponds to the partition $(w_1, w_2)(w_3, w_4)(w_5)$. It is clear that the circuit S is reachable. Using Remark 3.4, we obtain that in the integer model of computation, for the considered scheme of the circuit, the algorithm

$\mathscr{A}_2$ has the time complexity $O(n^3)$, which is polynomial depending on the length of the problem representation.

Let us consider two distinct cost functions $\Theta^{(1)}, \Theta^{(2)} \in \{\Psi^{(t)} : t = 1, 2, 3, 4\}$. We can use the algorithm $\mathscr{A}_4$ to find the set of Pareto optimal points for the problem of bi-criteria optimization of partitions relative to $\Theta^{(1)}$ and $\Theta^{(2)}$.

We denote $M = M_{\Theta^{(1)},\Theta^{(2)}}(S)$, where

$$M_{\Theta^{(1)},\Theta^{(2)}}(S) = \max\{|Par(t_{\Theta^{(1)},\Theta^{(2)}}(S, v))| : v \in V(S)\} .$$

To evaluate the number of operations from B^+ made by the algorithm $\mathscr{A}_4$, we use Theorem 4.1. To treat the input node of the scheme of S, we do not need any operations.

The scheme of S contains at most n^2 functional nodes. Each node has one entering edge. To process one functional node, the algorithm $\mathscr{A}_4$ makes at most $8M$ operations from B and at most $4M \log_2 M$ comparisons. Therefore, the algorithm needs at most $12n^2 M \log_2(2M)$ operations from B^+ to process all functional nodes.

For each of at most n unifying nodes with at most n entering edges, the algorithm $\mathscr{A}_4$ makes at most $4nM \log_2(nM)$ comparisons. Then, to process all unifying nodes, it needs at most $4n^2 M \log_2(nM)$ comparisons.

Hence, the algorithm $\mathscr{A}_4$ makes at most $16n^2 M \log_2(2nM)$ operations from B^+. Let $\Theta^{(1)} = \Psi^{(k)}$ and $\Theta^{(2)} = \Psi^{(l)}$ for $k, l \in \{1, 2, 3, 4\}$, $k \neq l$. By Proposition 4.4, $M \leq \rho_k n + 1$. Then, the algorithm $\mathscr{A}_4$ makes at most

$$16n^2(\rho_k n + 1) \log_2(2n(\rho_k n + 1))$$

operations from B^+.

We know that $\max(N_{\Psi^{(k)}}(S), N_{\Psi^{(l)}}(S)) \leq L^3 n$. Using Remark 4.4 we obtain that, for the algorithm $\mathscr{A}_4$, the number $L^3 n$ is an upper bound on the absolute values of the numbers at the inputs of the operations $\psi_v^{(k)}, \psi_v^{(l)}$ related to the functional nodes and at the inputs of comparisons related to the functional and unifying nodes of S. It is easy to check that, for the algorithm $\mathscr{A}_4$, the number $L^3 n$ is an upper bound on the absolute values of the numbers at the inputs of operations from B in formulas for the functions $\psi_v^{(k)}$ and $\psi_v^{(l)}$ (note that we do not need to recalculate already known values $l(w_{j+1}, \dots, w_i)$, and all these values used by the algorithm $\mathscr{A}_4$ are at most L).

Let $k = 1$. Then, the algorithm $\mathscr{A}_4$ makes at most $16n^2(L^2 n + 1) \log_2(2n(L^2 n + 1))$ operations from B^+, i.e., pseudo-polynomial number of operations (polynomial depending on the dimension n and the numeric parameter L). In the software model of computation, for the considered scheme of the circuit and pair of cost functions, the algorithm $\mathscr{A}_4$ has the time complexity $O(n^3 L^2 \log(nL))$, which is pseudo-polynomial (polynomial depending on the dimension n and the numeric parameter L). In the integer model of computation, for the considered circuit and pair of cost functions, the algorithm $\mathscr{A}_4$ has the time complexity

$$O(n^3 L^2 \log(nL)^2 \log\log(nL) \log\log\log(nL)) ,$$

which is pseudo-polynomial (polynomial depending on the length of the problem representation and the numeric parameter L).

Let $k = 2$. Then, the algorithm $\mathscr{A}_4$ makes at most $16n^2(L^3n + 1)\log_2(2n(L^3n + 1))$ operations from B^+, i.e., pseudo-polynomial number of operations (polynomial depending on the dimension n and the numeric parameter L). In the software model of computation, for the considered scheme of the circuit and pair of cost functions, the algorithm $\mathscr{A}_4$ has the time complexity $O(n^3L^3\log(nL))$, which is pseudo-polynomial (polynomial depending on the dimension n and the numeric parameter L). In the integer model of computation, for the considered circuit and pair of cost functions, the algorithm $\mathscr{A}_4$ has the time complexity

$$O(n^3L^3\log(nL)^2\log\log(nL)\log\log\log(nL)) ,$$

which is pseudo-polynomial (polynomial depending on the length of the problem representation and the numeric parameter L).

Let $k \in \{3, 4\}$. Then, the algorithm $\mathscr{A}_4$ makes at most $16n^2(n + 1)\log_2(2n(n + 1))$ operations from B^+, i.e., polynomial number of operations depending on the dimension n. In the software model of computation, for the considered scheme of the circuit and pair of cost functions, the algorithm $\mathscr{A}_4$ has the time complexity $O(n^3\log n)$, which is polynomial depending on the dimension n. In the integer model of computation, for the considered circuit and pair of cost functions, the algorithm $\mathscr{A}_4$ has the time complexity $O(n^3\log(nL)^2\log\log(nL)\log\log\log(nL))$, which is polynomial depending on the length of the problem representation.

10.3 Example

Let us consider the sequence of words w_1, w_2, w_3, w_4, w_5 from [1], where $w_1 = w_2 = w_3 = w_4 =$ blah and $w_5 =$ reallylongword. We assume that $L = 16$, $p_1 = 0$, $p_2 = 0$, $p_3 = 1$, $p_4 = 0$, and $p_5 = 0$.

We construct the circuit S_{LB}, which describes all partitions of this sequence—see Fig. 10.1. Apply to the circuit S_{LB} the procedure of optimization relative to $\Psi^{(1)}$. As a result, we have the circuit $S_{LB}^{\Psi^{(1)}}$, which can be obtained from the circuit S_{LB} by the removal of edges labeled with the number 1. The circuit $S_{LB}^{\Psi^{(1)}}$ describes the only partition $(w_1, w_2)(w_3, w_4)(w_5)$—see Fig. 10.2. We apply to the circuit $S_{LB}^{\Psi^{(1)}}$ the procedure of optimization relative to $\Psi^{(4)}$. As a result, we obtain the circuit $S_{LB}^{\Psi^{(1)},\Psi^{(4)}} = S_{LB}^{\Psi^{(1)}}$.

We also apply the procedure of bi-criteria optimization relative to $\Psi^{(1)}$ and $\Psi^{(4)}$ to the circuit S_{LB}. The set of Pareto optimal points for each node of S_{LB} is depicted in Fig. 10.3.

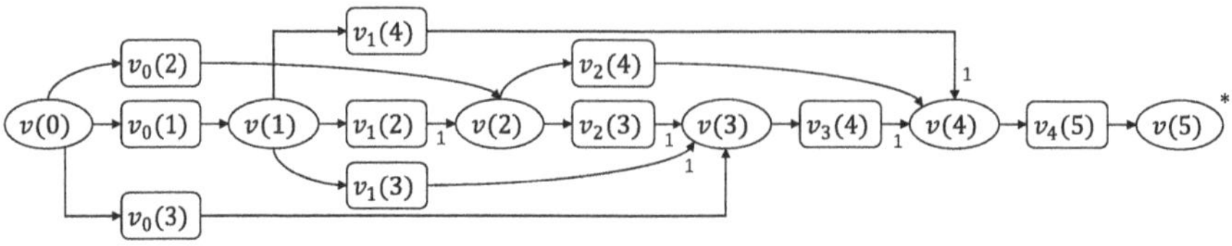

Fig. 10.1 Circuit S_{LB} for the sequence $w_1, \ldots, w_5$

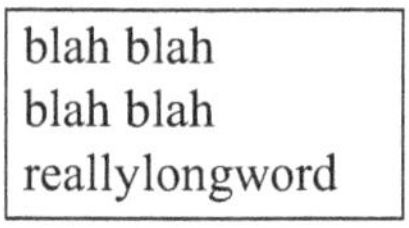

Fig. 10.2 Graphical representation of the partition $(w_1, w_2)(w_3, w_4)(w_5)$

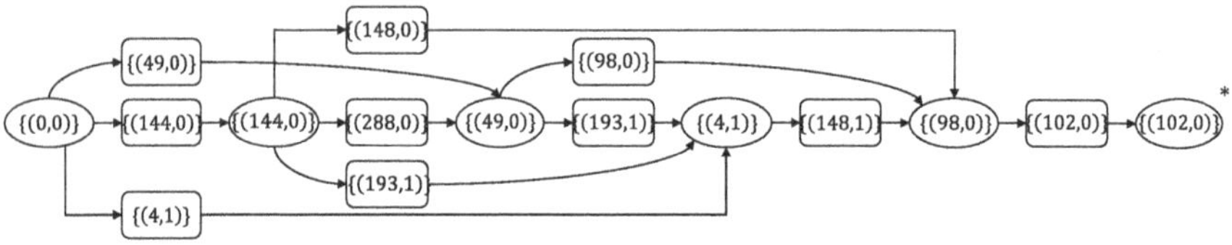

Fig. 10.3 Sets of Pareto optimal points for all nodes of the circuit S_{LB}

10.4 Experiments

We experimentally tested various optimization scenarios of partitioning a sequence of words into lines. Each experiment was repeated 20 times.

We randomly generated n lengths of words $|w_1|, \ldots, |w_n|$ from the range [2, 8] with uniform distribution. We generated penalties $p_1, \ldots, p_n \in \{0, 1\}$ such that the probability of $p_i = 1$ is equal to 0.2. A line has the length limit $L = 20, 30, 40, 50$.

Table 10.1 shows for $n = 200, 250, \ldots, 500$ and $L = 30$, the average number of partitions described by the circuit before optimization, after optimization relative to $\Psi^{(1)}$, relative to $\Psi^{(2)}$, relative to $\Psi^{(3)}$, and relative to $\Psi^{(4)}$. The experiments resulted in a few numbers of optimal partitions relative to $\Psi^{(1)}$ and $\Psi^{(2)}$ (badness of the partition) but in many optimal partitions relative to $\Psi^{(3)}$ and $\Psi^{(4)}$ (the number of lines and line breaking penalty).

Table 10.2 shows for $n = 200, 250, \ldots, 500$ and $L = 30$, the average number of partitions described by the circuit after optimization relative to $\Psi^{(1)}, \Psi^{(3)}$, relative to $\Psi^{(3)}, \Psi^{(1)}$, relative to $\Psi^{(1)}, \Psi^{(4)}$, relative to $\Psi^{(4)}, \Psi^{(1)}$, relative to $\Psi^{(3)}, \Psi^{(4)}$, and relative to $\Psi^{(4)}, \Psi^{(3)}$. The average number of partitions relative to $\Psi^{(1)}, \Psi^{(3)}$ and relative to $\Psi^{(3)}, \Psi^{(1)}$ was identical. Optimizing the circuit relative to $\Psi^{(1)}$ during the first or second optimization stage resulted in at most 59.2 optimal partitions on average.

Table 10.3 shows the average value of $\Psi^{(1)}$ after optimization relative to $\Psi^{(1)}$, relative to $\Psi^{(2)}, \Psi^{(1)}$, relative to $\Psi^{(3)}, \Psi^{(1)}$, and relative to $\Psi^{(4)}, \Psi^{(1)}$, average value

Table 10.1 Average number of partitions described by the circuit before optimization, after optimization relative to $\Psi^{(1)}$, relative to $\Psi^{(2)}$, relative to $\Psi^{(3)}$, and relative to $\Psi^{(4)}$ with $L = 30$ among 20 trials

n	$\emptyset$	$\Psi^{(1)}$	$\Psi^{(2)}$	$\Psi^{(3)}$	$\Psi^{(4)}$
200	2.52×10^{58}	10.10	8.40	1.03×10^{08}	6.88×10^{44}
250	2.11×10^{73}	9.40	9.10	1.21×10^{09}	8.28×10^{55}
300	1.37×10^{88}	9.35	7.55	1.12×10^{11}	2.01×10^{67}
350	6.61×10^{102}	10.40	10.60	1.42×10^{15}	7.96×10^{78}
400	3.89×10^{117}	30.65	32.80	1.59×10^{15}	1.49×10^{90}
450	1.01×10^{132}	27.15	23.00	6.13×10^{17}	5.44×10^{101}
500	4.43×10^{146}	38.90	35.25	7.33×10^{18}	6.53×10^{112}

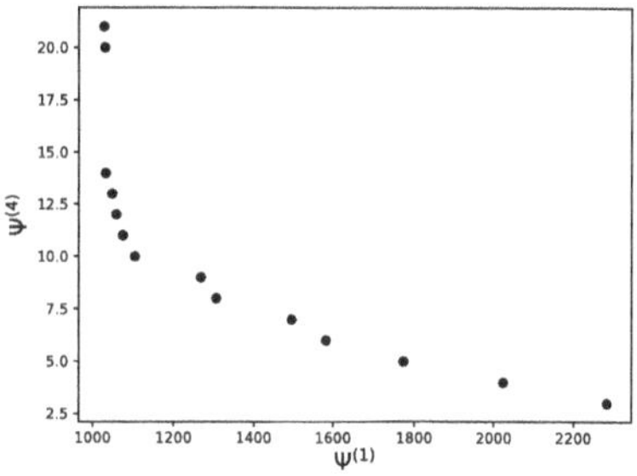

Fig. 10.4 Set of Pareto optimal points for a bi-criteria optimization of word sequence partition ($n = 500$, $L = 30$) relative to $\Psi^{(1)}$ and $\Psi^{(4)}$

of $\Psi^{(3)}$ after optimization relative to $\Psi^{(3)}$ and relative to $\Psi^{(4)}$, $\Psi^{(3)}$, and average value of $\Psi^{(4)}$ after optimization relative to $\Psi^{(4)}$ and relative to $\Psi^{(3)}$, $\Psi^{(4)}$. The experiments resulted in a significant increase in the value of $\Psi^{(4)}$ while comparing optimization relative to $\Psi^{(4)}$ and relative to $\Psi^{(3)}$, $\Psi^{(4)}$. A similar situation is with the optimization relative to $\Psi^{(1)}$ and relative to $\Psi^{(4)}$, $\Psi^{(1)}$.

We also consider bi-criteria optimization of partitions. Table 10.4 shows the minimum, maximum, and the average number of Pareto optimal points for bi-criteria optimization problems relative to $\Psi^{(1)}$ and $\Psi^{(4)}$ for $n = 200, 250, \ldots, 500$ and $L = 20, 30, 40, 50$. The minimum number of Pareto optimal points was two and the maximum number was 34 among all tested scenarios. None of the partitions were totally optimal relative to $\Psi^{(1)}$ and $\Psi^{(4)}$—see Table 10.6. The set of Pareto optimal points for a bi-criteria optimization of partitions ($n = 500$, $L = 30$) relative to $\Psi^{(1)}$ and $\Psi^{(4)}$ is depicted in Fig. 10.4.

Table 10.5 shows the minimum, maximum, and the average number of Pareto optimal points for bi-criteria optimization problem relative to $\Psi^{(1)}$, $\Psi^{(3)}$ for $n = 200, 250, \ldots, 500$ and $L = 20, 30, 40, 50$. The number of Pareto optimal points was one in all tested scenarios. There was a totally optimal partition relative to $\Psi^{(1)}$, $\Psi^{(3)}$ for each of the considered sequences of words—see Table 10.7. The set of Pareto optimal points for a bi-criteria optimization of partitions ($n = 500$, $L = 30$) relative to $\Psi^{(1)}$ and $\Psi^{(3)}$ is depicted in Fig. 10.5.

Table 10.2 Average number of partitions described by the circuit after optimization relative to $\Psi^{(1)}, \Psi^{(3)}$, relative to $\Psi^{(3)}, \Psi^{(1)}$, relative to $\Psi^{(1)}, \Psi^{(4)}$, relative to $\Psi^{(4)}, \Psi^{(1)}$, relative to $\Psi^{(3)}, \Psi^{(4)}$, and relative to $\Psi^{(4)}, \Psi^{(3)}$ with $L = 30$ among 20 trials

n	$\Psi^{(1)}, \Psi^{(3)}$	$\Psi^{(3)}, \Psi^{(1)}$	$\Psi^{(1)}, \Psi^{(4)}$	$\Psi^{(4)}, \Psi^{(1)}$	$\Psi^{(3)}, \Psi^{(4)}$	$\Psi^{(4)}, \Psi^{(3)}$
200	18.00	18.00	4.10	3.50	9.84×10^{04}	4.10×10^{10}
250	13.50	13.50	6.20	3.20	4.57×10^{05}	4.49×10^{11}
300	13.30	13.30	5.60	7.40	5.99×10^{05}	5.85×10^{12}
350	13.40	13.40	7.10	8.10	9.65×10^{07}	6.55×10^{16}
400	53.60	53.60	9.10	5.50	3.04×10^{06}	7.26×10^{18}
450	37.80	37.80	7.90	9.00	1.33×10^{09}	4.69×10^{20}
500	59.20	59.20	16.10	37.10	3.11×10^{10}	2.17×10^{23}

Table 10.3 Average value of $\Psi^{(1)}$ after optimization relative to $\Psi^{(1)}$, relative to $\Psi^{(2)}, \Psi^{(1)}$, relative to $\Psi^{(3)}, \Psi^{(1)}$, and relative to $\Psi^{(4)}, \Psi^{(1)}$, average value of $\Psi^{(3)}$ after optimization relative to $\Psi^{(3)}$ and relative to $\Psi^{(4)}, \Psi^{(3)}$, and average value of $\Psi^{(4)}$ after optimization relative to $\Psi^{(4)}$ and relative to $\Psi^{(3)}, \Psi^{(4)}$ with $L = 30$ among 20 trials

n	$\Psi^{(1)}$	$\Psi^{(2)}, \Psi^{(1)}$	$\Psi^{(3)}, \Psi^{(1)}$	$\Psi^{(4)}, \Psi^{(1)}$	$\Psi^{(3)}$	$\Psi^{(4)}, \Psi^{(3)}$	$\Psi^{(4)}$	$\Psi^{(3)}, \Psi^{(4)}$
200	432.30	432.50	443.80	888.60	39.25	41.10	0.25	3.20
250	519.80	520.80	543.10	1.27×10^{03}	48.95	51.80	0.20	5.35
300	607.30	607.30	625.60	1.27×10^{03}	58.50	61.10	0.15	5.55
350	731.20	731.40	743.00	1.56×10^{03}	68.40	71.30	0.45	6.30
400	795.80	797.20	815.00	1.84×10^{03}	77.70	81.35	0.35	7.30
450	842.90	843.80	860.80	1.90×10^{03}	87.10	91.40	0.40	8.50
500	972.85	973.85	993.70	2.10×10^{03}	97.10	101.85	0.30	8.95

Table 10.4 Number of Pareto optimal points (in format ${}_{\min}average_{\max}$ among 20 trials) for bi-criteria optimization problem relative to $\Psi^{(1)}, \Psi^{(4)}$

n	L			
	20	30	40	50
200	${}_{6}11.00_{16}$	${}_{5}7.65_{12}$	${}_{3}5.95_{9}$	${}_{2}4.95_{8}$
250	${}_{8}14.15_{21}$	${}_{6}10.10_{15}$	${}_{4}6.45_{9}$	${}_{3}5.55_{8}$
300	${}_{9}15.80_{21}$	${}_{7}11.50_{17}$	${}_{4}8.05_{13}$	${}_{2}6.15_{9}$
350	${}_{14}19.00_{27}$	${}_{8}13.05_{18}$	${}_{5}9.60_{15}$	${}_{5}7.65_{11}$
400	${}_{16}21.05_{30}$	${}_{9}14.80_{21}$	${}_{9}11.35_{15}$	${}_{6}9.55_{16}$
450	${}_{18}23.35_{32}$	${}_{12}16.35_{20}$	${}_{8}12.45_{18}$	${}_{4}9.50_{13}$
500	${}_{20}25.10_{34}$	${}_{12}16.90_{22}$	${}_{10}13.30_{20}$	${}_{6}9.95_{13}$

Table 10.5 Number of Pareto optimal points (in format ${}_{\min}average_{\max}$ among 20 trials) for bi-criteria optimization problem relative to $\Psi^{(1)}$, $\Psi^{(3)}$

n	L			
	20	30	40	50
200	${}_1 1.00_1$	${}_1 1.00_1$	${}_1 1.00_1$	${}_1 1.00_1$
250	${}_1 1.00_1$	${}_1 1.00_1$	${}_1 1.00_1$	${}_1 1.00_1$
300	${}_1 1.00_1$	${}_1 1.00_1$	${}_1 1.00_1$	${}_1 1.00_1$
350	${}_1 1.00_1$	${}_1 1.00_1$	${}_1 1.00_1$	${}_1 1.00_1$
400	${}_1 1.00_1$	${}_1 1.00_1$	${}_1 1.00_1$	${}_1 1.00_1$
450	${}_1 1.00_1$	${}_1 1.00_1$	${}_1 1.00_1$	${}_1 1.00_1$
500	${}_1 1.00_1$	${}_1 1.00_1$	${}_1 1.00_1$	${}_1 1.00_1$

Table 10.6 Number of cases with totally optimal partitions relative to $\Psi^{(1)}$ and $\Psi^{(4)}$ among 20 trials

n	L			
	20	30	40	50
200	0	0	0	0
250	0	0	0	0
300	0	0	0	0
350	0	0	0	0
400	0	0	0	0
450	0	0	0	0
500	0	0	0	0

Table 10.7 Number of cases with totally optimal partitions relative to $\Psi^{(1)}$ and $\Psi^{(3)}$ among 20 trials

n	L			
	20	30	40	50
200	20	20	20	20
250	20	20	20	20
300	20	20	20	20
350	20	20	20	20
400	20	20	20	20
450	20	20	20	20
500	20	20	20	20

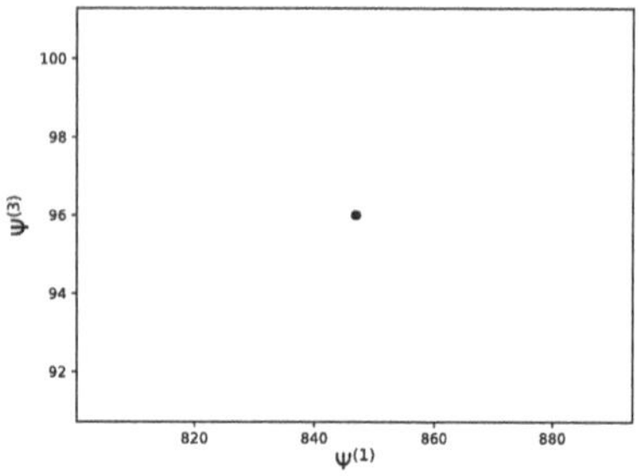

Fig. 10.5 Set of Pareto optimal points for a bi-criteria optimization of word sequence partition ($n = 500$, $L = 30$) relative to $\Psi^{(1)}$ and $\Psi^{(3)}$

References

1. Demaine, E., Devadas, S.: 6.006 Introduction to Algorithms. Fall 2011. Massachusetts Institute of Technology. MIT OpenCourseWare. https://ocw.mit.edu
2. Knuth, D.E., Plass, M.F.: Breaking paragraphs into lines. Softw. Pract. Exp. **11**(11), 1119–1184 (1981)
3. Mankowski, M., Moshkov, M.: Extensions of dynamic programming for multi-stage combinatorial optimization. Theor. Comput. Sci. **844**, 106–132 (2020)

Chapter 11
One-Dimensional Clustering

The clustering is a task of grouping objects, which are similar with respect to some metric. Clustering is a well-known problem in statistics, data mining, and machine learning. There are multiple definitions and approaches to the clustering (see position paper by Estivill-Castro [1]). Sometimes, clustering is a combinatorial optimization problem of finding a partition of a set with the minimum cost.

We consider one-dimensional k-clustering of points from $\mathbb{R}$ and three cost functions. The first one is the sum of squared Euclidean distances between points and means of clusters to which the points belong. The second one is the sum of distances between points and medians of corresponding clusters. The third cost function is the maximum length of a cluster. The dynamic programming algorithms for the optimization of k-clusterings for the first and the second cost functions were proposed in [4] and [2], respectively.

In this chapter, we design a conventional circuit without repetitions for the problem of one-dimensional clustering, describe three cost functions for this circuit, evaluate the number of operations and the time required by the algorithms for the optimization and counting of clusterings, discuss an example, and show experimental results. This chapter is partially based on [3].

11.1 Definition of Circuit and Cost Functions

We consider a sequence $a_1 < a_2 < \cdots < a_n$ of $n \geq 1$ nonnegative real numbers. For natural k, $1 \leq k \leq n$, k-clustering is a partition of this sequence into k nonempty, non-overlapping, and contiguous clusters $[a_1, \ldots, a_{p-1}][a_p, \ldots, a_{p+t}] \cdots [a_q, \ldots, a_n]$.

The numbers n and k are the dimensions of the problems under consideration related to the optimization of clusterings. To describe an instance of a problem, we need to specify n numbers $a_1, \ldots, a_n$ and the number k. Note that n and k are at most the number of numeric parameters of the problem.

M. Mankowski and M. Moshkov, *Dynamic Programming Multi-Objective Combinatorial Optimization*, Studies in Systems, Decision and Control 331,
https://doi.org/10.1007/978-3-030-63920-4_11

Let us assume that $a_1, \ldots, a_n$ are integers. The binary representations of these numbers and the number k form the representation of the problem instance (we will omit the word instance). It is clear that n and k are at most the length of the problem representation.

We denote by $\alpha \otimes \beta$ the concatenation $\alpha\beta$ of sequences of clusters α and β. For a set A of sequences of clusters, we denote $A \otimes \beta = \{\alpha \otimes \beta : \alpha \in A\}$.

For $1 \leq m \leq k$ and $m \leq i \leq n-k+m$, we denote by $T(i, m)$ the set of all m-clusterings of the sequence $a_1, a_2, \ldots, a_i$. Then $T(i, 1) = \{[a_1, \ldots, a_i]\}$ for $i = 1, 2, \ldots, n-k+1$. For $1 < m \leq k$ and $m \leq i \leq n-k+m$, we have

$$T(i, m) = \bigcup_{m \leq j \leq i} T(j-1, m-1) \otimes [a_j, \ldots, a_i] .$$

Set

$$U = T(n, k) \cup \bigcup_{\substack{1 \leq m < k \\ m \leq i \leq n-k+m}} T(i, m) .$$

We define a circuit S without repetitions over the set U, which describes the set $T(n, k)$ of all k-clusterings of the sequence $a_1, \ldots, a_n$. Let $\mu_{i,j,m}$, $1 < m \leq k$ and $m \leq j \leq i \leq n-k+m$, be a concatenation function from $T(j-1, m-1)$ to $T(i, m)$ such that $\mu_{i,j,m}(x) = x \otimes [a_j, \ldots, a_i]$ for any $x \in T(j-1, m-1)$.

For $i = 1, \ldots, n-k+1$, the circuit S contains an input node $v(i, 1)$ labeled with the set $\{[a_1, \ldots, a_i]\}$. For each m, i, $1 < m < k$ and $m \leq i \leq n-k+m$, the circuit S contains a unifying node $v(i, m)$ labeled with $\cup$. There is also a unifying node $v(n, k)$ labeled with $\cup$ and $*$, which is considered as the output of S. For each m, j, i such that $1 < m < k$, $m \leq i \leq n-k+m$, and $m \leq j \leq i$, the circuit S contains a functional node $v(i, j, m)$ labeled with the function $\mu_{i,j,m}$. For each j, $k \leq j \leq n$, the circuit S contains a functional node $v(n, j, k)$ labeled with the function $\mu_{k,j,n}$.

The unifying node $v(p, q)$, $v(p, q) \in V_u(S)$, has $p-q+1$ entering edges leaving the functional nodes $v(p, j, q)$, $p \leq j \leq q$. The functional node $v(p, j, q)$, $v(p, j, q) \in V_f(S)$, has one entering edge leaving $v(j-1, q-1)$. We will not label this edge with the number 1.

One can show that $S(v(i, m)) = T(i, m)$ for $1 \leq m < k, m \leq i \leq n-k+m$, and $S(v(i, j, m)) = T(j-1, m-1) \otimes [a_j, \ldots, a_i]$ for any functional node $v(i, j, m)$ of S. Also, $S(v(n, k)) = T(n, k)$.

The described circuit will be considered as the preliminary version of the circuit S. The final version of S is obtained from the preliminary version of S by the removal of all nodes such that there is no directed path from the considered node to the output. We also remove all edges that are incident to the removed nodes. Later, S without mentioning the version denotes the final version of the circuit S.

The preliminary version of the circuit S contains at most n input nodes, at most nk unifying nodes with at most n entering edges each, and at most n^2k functional nodes with one entering edge each. Hence, it contains at most $n + nk + n^2k$ nodes and at most $2n^2k$ edges.

Let us prove that S is a circuit without repetitions. It is clear that the concatenation functions $\mu_{i,j,m}$ are injective. Let v_1 and v_2 be different nodes from $V_i(S) \cup V_f(S)$. We now show that $S(v_1) \cap S(v_2) = \emptyset$. Let v_1 and v_2 be any two different input nodes from $V_i(S)$. Then, for some i_1 and i_2, $i_1 \neq i_2$, $S(v_1) = \{[a_1, \dots, a_{i_1}]\}$ and $S(v_2) = \{[a_1, \dots, a_{i_2}]\}$. It is clear that $S(v_1) \cap S(v_2) = \emptyset$. Let v_1 be an input node and v_2 be a functional node. It is clear that sets $S(v_1)$ and $S(v_2)$ do not contain common elements since $S(v_1)$ contains only one clustering with one cluster (1-clustering) but $S(v_2)$ does not contain any 1-clustering. Therefore, $S(v_1) \cap S(v_2) = \emptyset$. Let v_1 and v_2 be functional nodes and $v_1 = v(i_1, j_1, m_1)$, $v_2 = v(i_2, j_2, m_2)$. If $i_1 \neq i_2$, then $S(v(i_1, j_1, m_1))$ and $S(v(i_2, j_2, m_2))$ describe clusterings of sequences with different lengths. If $m_1 \neq m_2$, then $S(v(i_1, j_1, m_1))$ and $S(v(i_2, j_2, m_2))$ describe clusterings with a different number of clusters. If $j_1 \neq j_2$, then $S(v(i_1, j_1, m_1))$ and $S(v(i_2, j_2, m_2))$ describe clusterings, where the last cluster in each clusterings starts with either a_{j_1} or a_{j_2}. Thus, $S(v_1) \cap S(v_2) = \emptyset$.

We now define a cost function $\Psi^{(t)}$, $t = 1, 2, 3$, of the circuit S given by the functions $\psi^{(t)}$ and $\psi_v^{(t)}$, $v \in V_f(S)$. Denote

$$B = \{x + y, x - y, x \cdot y, x/y, \max(x, y), \lfloor x \rfloor, \lceil x \rceil, |x|\} .$$

The k-means clustering is a clustering problem with the objective of minimizing the sum of squared distances between points and means of corresponding clusters. We define the mean $\zeta_{j,i}$ of a cluster $[a_j, \dots, a_i]$ such that $\zeta_{j,i} = \Sigma_{p=j}^{i} a_p/(i - j + 1)$.

The cost function $\Psi^{(1)}$ is defined by the functions $\psi^{(1)}$ and $\psi^{(1)}_{v(i,j,m)}$, $v(i, j, m) \in V_f(S)$, where $\psi^{(1)}([a_1, \dots, a_i]) = \sum_{p=1}^{i} (a_p - \zeta_{1,i})^2$ for $i = 1, 2, \dots, n - k + 1$, and $\psi^{(1)}_{v(i,j,m)}(x) = x + \sum_{p=j}^{i} (a_p - \zeta_{j,i})^2$. Note that

$$\begin{aligned} \sum_{p=j}^{i} (a_p - \zeta_{j,i})^2 &= \sum_{p=j}^{i} (a_p^2 - 2a_p\zeta_{j,i} + \zeta_{j,i}^2) \\ &= (i - j + 1)\zeta_{j,i}^2 + \sum_{p=j}^{i} a_p^2 - 2\zeta_{j,i} \sum_{p=j}^{i} a_p . \end{aligned}$$

The mean $\zeta_{j,i}$ and the sum $\sum_{p=j}^{i} (a_p - \zeta_{j,i})^2$ can be calculated using the prefix sum sequence $P_1 = P_1(0), P_1(1), \dots, P_1(n)$, where $P_1(0) = 0$ and $P_1(b) = P_1(b - 1) + a_b$ for $b = 1, \dots, n$, and the prefix sum sequence $P_2 = P_2(0), P_2(1), \dots, P_2(n)$, where $P_2(0) = 0$ and $P_2(b) = P_2(b - 1) + a_b^2$ for $b = 1, \dots, n$. To compute the sequence P_1, we need n operations of addition, and to compute the sequence P_2, we need $2n$ operations of addition and multiplication. To derive $\zeta_{j,i}$ from P_1, we need four operations from B: $\zeta_{j,i} = (P_1(i) - P_1(j - 1))/(i - j + 1)$. To derive $\sum_{p=j}^{i} (a_p - \zeta_{j,i})^2$ from $\zeta_{j,i}$, P_1, and P_2 we need 10 operations from B: $\sum_{p=j}^{i} (a_p - \zeta_{j,i})^2 = (i - j + 1)\zeta_{j,i}^2 + (P_2(i) - P_2(j - 1)) - 2\zeta_{j,i}(P_1(i) - P_1(j - 1))$. Therefore, to compute the value of $\psi^{(1)}_{v(i,j,m)}$, knowing P_1 and P_2, we need 15 operations from the set B. We can show in a similar way that, to compute the value of $\psi^{(1)}$, knowing P_1 and P_2, we need seven operations from B. The cost function $\Psi^{(1)}$ is strictly increasing. This function is not an integer cost function.

The k-medians clustering is a clustering problem with the objective to minimize the sum of distances between points and medians of corresponding clusters. We define the median $\eta_{j,i}$ of a cluster $[a_j, \dots, a_i]$ such that $\eta_{i,j} = (a_{\lfloor m_{j,i} \rfloor} + a_{\lceil m_{j,i} \rceil})/2$, where $m_{j,i} = (i + j)/2$. The cost function $\Psi^{(2)}$ is defined by the functions $\psi^{(2)}$ and $\psi^{(2)}_{v(i,j,m)}$, $v(i,j,m) \in V_f(S)$, where $\psi^{(2)}([a_1, \dots, a_i]) = \sum_{p=1}^{i} |a_p - \eta_{1,i}|$ for $i = 1, 2, \dots, n - k + 1$ and $\psi^{(2)}_{v(i,j,m)}(x) = x + \sum_{p=j}^{i} |a_p - \eta_{j,i}|$. As in [2],

$$
\begin{aligned}
\sum_{p=j}^{i} |a_p - \eta_{j,i}| &= \sum_{p=j}^{\lfloor m_{j,i} \rfloor} (\eta_{j,i} - a_p) + \sum_{p=1+\lfloor m_{j,i} \rfloor}^{i} (a_p - \eta_{j,i}) \\
&= \sum_{p=1+\lfloor m_{j,i} \rfloor}^{i} a_p - \sum_{p=j}^{\lfloor m_{j,i} \rfloor} a_p + \eta_{j,i}(2\lfloor m_{j,i} \rfloor - i - j + 1) \, .
\end{aligned}
$$

To compute $\lfloor m_{j,i} \rfloor$ and $\eta_{j,i}$, we need six operations from B. To derive $\sum_{p=j}^{i} |a_p - \eta_{j,i}|$ from P_1, $\lfloor m_{j,i} \rfloor$, and $\eta_{j,i}$, we need nine operations from B: $\sum_{p=j}^{i} |a_p - \eta_{j,i}| = (P_1(i) - P_1(\lfloor m_{j,i} \rfloor)) - (P_1(\lfloor m_{j,i} \rfloor) - P_1(j-1)) + \eta_{j,i}(2\lfloor m_{j,i} \rfloor - i - j + 1)$.

To compute $\psi^{(2)}_{v(i,j,m)}$ knowing prefix sum sequence P_1, we need 16 operations from B. We can show in a similar way that, to compute the value of $\psi^{(2)}$, knowing P_1, we need 12 operations from B. The cost function $\Psi^{(2)}$ is strictly increasing. This function is not an integer cost function.

We also consider the minimization of the maximum length of a cluster. The length of the cluster $[a_j, \dots, a_i]$ is $a_i - a_j$. The cost function $\Psi^{(3)}$ is defined by the functions $\psi^{(3)}$ and $\psi^{(3)}_{v(i,j,m)}$, $v(i,j,m) \in V_f(S)$, where $\psi^{(3)}([a_1, \dots, a_i]) = a_i - a_1$ for $i = 1, 2, \dots, n$ and $\psi^{(3)}_{v(i,j,m)}(x) = \max(x, a_i - a_j)$. To compute $\psi^{(3)}$, we need one operation from B, and to compute $\psi^{(3)}_{v(i,j,m)}$ we need two operations from B. The cost function $\Psi^{(3)}$ is an increasing nonnegative integer cost function.

All these cost functions are adapted (see Sect. 2.4). They are described by formulas over the set of elementary operations B.

11.2 Time Complexity of Algorithms

The algorithms $\mathscr{A}_1$, $\mathscr{A}_2$, and $\mathscr{A}_4$ work with the scheme of the circuit S (see Sect. 2.4). The preliminary version of the circuit S is completely described. Therefore, the time complexity of the scheme of the preliminary version of S construction is linear depending on the number of its nodes and edges, i.e., $O(n^2k)$ (or $O(n^3)$

since $k \leq n$). Using breadth-first search from the output node in the scheme of the preliminary version of S with the reverse direction of the edges, we can find in the scheme of the preliminary version of S all nodes such that there is no directed path from the considered node to the output and construct the scheme of S in $O(n^3)$ time. In total, we need $O(n^3)$ time to construct the scheme of S.

Note that the algorithms $\mathscr{A}_1$ and $\mathscr{A}_2$ can also work with the schemes of edge-preserving subcircuits of S obtained as a result of the algorithm $\mathscr{A}_1$ work—see Remark 3.1.

Denote $B^+ = B \cup \{x : y\}$, where $x : y$ is the operation of comparison with values $x < y$, $x = y$, and $x > y$.

Before the use of the algorithms $\mathscr{A}_1$ and $\mathscr{A}_4$ with at least one of the cost functions $\Psi^{(1)}, \Psi^{(2)}$, we compute the prefix sum sequences P_1 and P_2, which requires $3n$ operations of addition and multiplication. In the software model of computation, this step requires $O(n)$ time.

We can use the algorithm $\mathscr{A}_1$ for multi-stage optimization of clusterings. Using Theorem 3.1, we evaluate the number of operations required by the algorithm $\mathscr{A}_1$ working with the scheme of an edge-preserving subcircuit T of the circuit S and a cost function $\Psi^{(t)}$ for S, where $t \in \{1, 2, 3\}$.

The scheme of the circuit T contains at most n input nodes. To process one input node, we need at most 12 operations from B. Therefore, we need at most $12n$ operations from the set B^+ to treat all input nodes.

The scheme of T contains at most nk unifying nodes. Each unifying node has at most n entering edges. To treat all unifying nodes, the algorithm $\mathscr{A}_1$ makes at most $(2n - 1)nk \leq 2n^2k$ comparisons.

The scheme of T contains at most n^2k functional nodes. For each functional node v, the algorithm makes at most 16 operations from B to compute $\psi_v^{(t)}$. The number of operations to treat all functional nodes is at most $16n^2k$. The total number of operations from the set B^+ required by $\mathscr{A}_1$ is at most $30n^2k$, i.e., it is polynomial depending on the dimensions n and k.

In the software model of computation, for the considered scheme of the circuit and cost function, the algorithm $\mathscr{A}_1$ has polynomial time complexity $O(n^2k)$ depending on the problem dimensions n and k.

Let $t = 3$ and $a_1, \ldots, a_n$ be integers. One can show that $N_{\Psi^{(3)}}(S) \leq a_n$, where $N_{\Psi^{(3)}}(S) = \max\{|\Psi^{(3)}(\delta)| : \delta \in U(S)\}$ and $U(S) = \bigcup_{v \in V(S)} S(v)$. Note that $\log_2 a_n$ is at most the length of the problem representation. Using Remark 3.2 we obtain that, for the algorithm $\mathscr{A}_1$, the number a_n is an upper bound on the absolute values of the numbers at the inputs of the operations $\psi_v^{(3)}$ related to the functional nodes $v \in V_f(S)$ and at the inputs of comparisons related to the unifying nodes $v \in V_u(S)$. It is easy to check that, for the algorithm $\mathscr{A}_1$, the number a_n is an upper bound on the absolute values of the numbers at the inputs of operations from B in formulas for the functions $\psi_v^{(3)}$ and in formulas for the function $\psi^{(3)}$. The algorithm $\mathscr{A}_1$ uses only the operations of comparison and subtraction. Therefore, in the integer model of computation, for the considered scheme of the circuit and the cost function $\Psi^{(3)}$, the algorithm $\mathscr{A}_1$ has the time complexity $O(n^2k \log a_n)$, which is polynomial depending on the length of the problem representation (n, k, and $\log_2 a_n$ are at most the length of the problem representation).

To count all possible clusterings described by the circuit T, we apply the algorithm $\mathcal{A}_2$. We use Theorem 3.2 to evaluate the number of operations required by $\mathcal{A}_2$. The scheme of T contains at most nk unifying nodes, each with at most n entering edges. Therefore, we need at most n^2k additions to treat all unifying nodes. For each of at most n^2k functional nodes, and each of at most n input nodes, we do not need any operations. The total number of operations from B^+ (additions) is at most n^2k, i.e., it is polynomial depending on the dimensions n and k.

In the software model of computation, for the considered scheme of the circuit, the algorithm $\mathcal{A}_2$ has polynomial time complexity $O(n^2k)$ depending on the problem dimensions n and k.

We now evaluate the number $|S(v(n,k))|$, which is equal to the number of all k-clusterings of the sequence $a_1, \dots, a_n$. One can show that this value is at most 2^{n-1}: there is a natural one-to-one mapping of all clusterings (not only k-clusterings) to tuples from $\{0,1\}^n$ in which the first digit is equal to 1. For example, the tuple $(1,1,0,0,1)$ corresponds to the clustering $[a_1,a_2][a_3,a_4][a_5]$. It is clear that the circuit S is reachable. Using Remark 3.4, we obtain that in the integer model of computation, for the considered scheme of the circuit, the algorithm $\mathcal{A}_2$ has time complexity $O(n^3k)$, which is polynomial depending on the length of the problem representation.

We consider two distinct cost functions $\Theta^{(1)} = \Psi^{(3)}$ and $\Theta^{(2)} \in \{\Psi^{(t)} : t = 1,2\}$. We can use the algorithm $\mathcal{A}_4$ to find the set of Pareto optimal points for the problem of bi-criteria optimization of clusterings relative to $\Theta^{(1)}$ and $\Theta^{(2)}$.

We denote $M = M_{\Theta^{(1)},\Theta^{(2)}}(S)$, where

$$M_{\Theta^{(1)},\Theta^{(2)}}(S) = \max\{|Par(t_{\Theta^{(1)},\Theta^{(2)}}(S,v))| : v \in V(S)\} .$$

To evaluate the number of operations required by the algorithm $\mathcal{A}_4$, we use Theorem 4.1. To treat at most n input nodes of the scheme of S, we need at most $13n$ operations from B.

The scheme of the circuit S contains at most n^2k functional nodes. Each functional node has one entering edge. To process all functional nodes, the algorithm $\mathcal{A}_4$ makes at most $18n^2kM$ operations from B and at most $4n^2kM\log_2 M$ comparisons. Therefore, the algorithm requires at most $22n^2kM\log_2(2M)$ operations from B^+ to treat all functional nodes.

For each of nk unifying nodes with at most n entering edges, the algorithm $\mathcal{A}_4$ makes at most $4nkM\log_2(nM)$ comparisons. Then, to process all unifying nodes, we need at most $4n^2kM\log_2(nM)$ comparisons.

Hence, the algorithm $\mathcal{A}_4$ makes at most $39n^2kM\log_2(2nM)$ operations from the set B^+.

It is clear that $|\{\Psi^{(3)}(\delta) : \delta \in U(S)\}| \le n^2$: the length of a cluster can have at most n^2 different values. By Proposition 4.5, $M_{\Theta^{(1)},\Theta^{(2)}}(S) \le n^2$. Since $M \le n^2$, the algorithm $\mathcal{A}_4$ makes at most $39n^4k\log_2(2n^3)$ operations from B^+, i.e., polynomial number of operations depending on the dimensions n and k.

In the software model of computation, for the considered scheme of the circuit and pair of cost functions, the algorithm $\mathscr{A}_4$ has the time complexity $O(n^4 k \log n)$, which is polynomial depending on the dimensions n and k.

We have only one integer cost function $\Psi^{(3)}$. Therefore, we cannot consider the integer model of computation in the case of the algorithm $\mathscr{A}_4$.

11.3 Example

Let us consider the problem of the 3-clustering of the sequence $a_1, \ldots, a_5$, where $a_1 = 0, a_2 = 1, a_3 = 2, a_4 = 4, a_5 = 5$ (see Fig. 11.1).

We construct the circuit S_{1DC}, which describes all 3-clusterings of this sequence (see Fig. 11.2). Apply to the circuit S_{1DC} the procedure of optimization relative to $\Psi^{(2)}$ (k-medians). As a result, we have the circuit $S_{1DC}^{\Psi^{(2)}}$, which can be obtained from the circuit S_{1DC} by the removal of edges labeled with the number 1. The circuit $S_{1DC}^{\Psi^{(2)}}$ describes three clusterings:

$$[a_1][a_2, a_3][a_4, a_5] ,$$
$$[a_1, a_2][a_3][a_4, a_5] ,$$
$$[a_1, a_2, a_3][a_4][a_5] .$$

We apply to the circuit $S_{1DC}^{\Psi^{(2)}}$ the procedure of optimization relative to $\Psi^{(3)}$. As a result, we have the circuit $S_{1DC}^{\Psi^{(2)},\Psi^{(3)}}$, which can be obtained from the circuit S_{1DC} by the removal of edges labeled with the numbers 1 and 2 (see Fig. 11.2). The circuit $S_{1DC}^{\Psi^{(2)},\Psi^{(3)}}$ describes two clusterings:

$$[a_1][a_2, a_3][a_4, a_5] ,$$
$$[a_1, a_2][a_3][a_4, a_5] .$$

We also apply the procedure of bi-criteria optimization relative to $\Psi^{(2)}$ and $\Psi^{(3)}$ to the circuit S_{1DC}. For each node of S_{1DC}, the set of Pareto optimal points is depicted in Fig. 11.3.

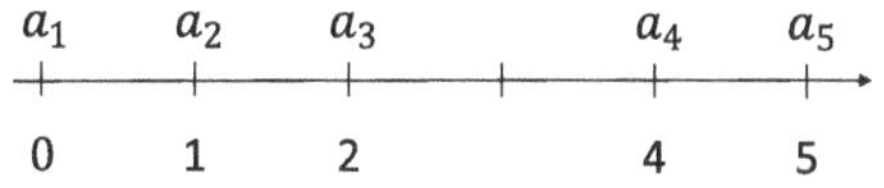

Fig. 11.1 Sequence $a_1, \ldots, a_5$

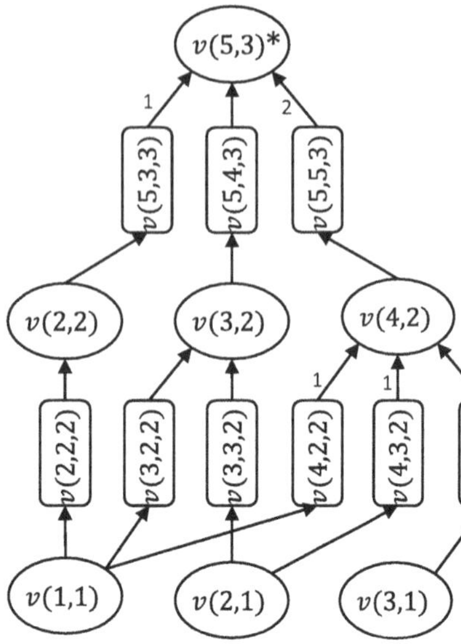

Fig. 11.2 Circuit S_{1DC} for the 3-clustering of the sequence $a_1, \ldots, a_5$

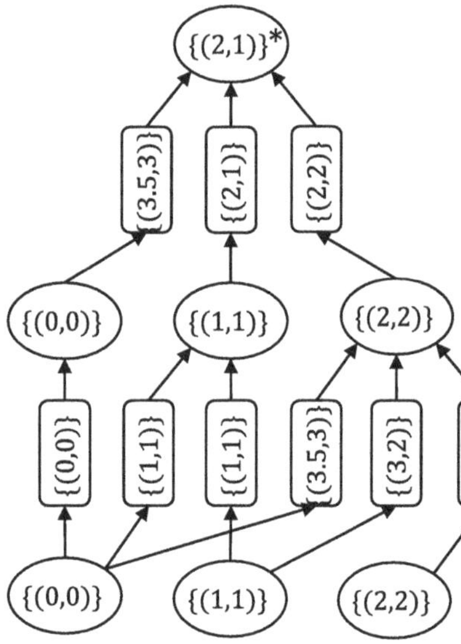

Fig. 11.3 Sets of Pareto optimal points for all nodes of the circuit S_{1DC}

11.4 Experiments

We experimentally tested different optimization scenarios for k-clustering of sorted sequences of integers. Each experiment was repeated 100 times. Each number in the sequence was randomly selected from the interval [0, 100] with uniform distribution.

Table 11.1 shows, for $n = 50, 100, \ldots, 350$, the average number of 5-clusterings described by the circuit before optimization, after optimization relative to $\Psi^{(1)}$, rela-

Table 11.1 Average number of 5-clusterings described by the circuit before optimization, after optimization relative to $\Psi^{(1)}$, relative to $\Psi^{(2)}$, and relative to $\Psi^{(3)}$ among 100 trials; points from [0, 100]

n	$\emptyset$	$\Psi^{(1)}$	$\Psi^{(2)}$	$\Psi^{(3)}$
50	2.12×10^{05}	1	2.07	5.40
100	3.76×10^{06}	1	3.20	11.80
150	1.97×10^{07}	1	5.62	12.70
200	6.34×10^{07}	1	6.49	8.35
250	1.56×10^{08}	1	8.05	7.24
300	3.26×10^{08}	1	11.90	5.10
350	6.08×10^{08}	1	17.73	3.58

Table 11.2 Average value of $\Psi^{(1)}$ after optimization relative to $\Psi^{(1)}$ and relative to $\Psi^{(2)}, \Psi^{(1)}$, average value of $\Psi^{(2)}$ after optimization relative to $\Psi^{(2)}$ and relative to $\Psi^{(1)}, \Psi^{(2)}$, and average value of $\Psi^{(3)}$ after optimization relative to $\Psi^{(3)}$, relative to $\Psi^{(1)}, \Psi^{(3)}$, and relative to $\Psi^{(2)}, \Psi^{(3)}$ among 100 trials, points from [0, 100], 5-clustering

n	$\Psi^{(1)}$	$\Psi^{(2)}, \Psi^{(1)}$	$\Psi^{(2)}$	$\Psi^{(1)}, \Psi^{(2)}$	$\Psi^{(3)}$	$\Psi^{(1)}, \Psi^{(3)}$	$\Psi^{(2)}, \Psi^{(3)}$
50	1.30×10^{03}	1.33×10^{03}	205.33	207.84	17.64	19.56	20.51
100	2.94×10^{03}	2.99×10^{03}	448.32	452.99	18.69	20.72	21.95
150	4.56×10^{03}	4.63×10^{03}	691.85	698.22	18.94	20.89	21.83
200	6.24×10^{03}	6.32×10^{03}	940.91	948.23	18.99	21.17	22.08
250	7.87×10^{03}	7.97×10^{03}	1.19×10^{03}	1.20×10^{03}	19.00	21.15	22.05
300	9.56×10^{03}	9.68×10^{03}	1.44×10^{03}	1.45×10^{03}	19.00	21.06	21.92
350	1.12×10^{04}	1.13×10^{04}	1.69×10^{03}	1.70×10^{03}	19.00	20.99	21.67

tive to $\Psi^{(2)}$, and relative to $\Psi^{(3)}$. Among tested scenarios, there was always one optimal clustering relative to $\Psi^{(1)}$, while there were at least 2.07 optimal 5-clusterings relative to $\Psi^{(2)}$ and at least 3.58 optimal 5-clusterings relative to $\Psi^{(3)}$ on average.

Table 11.2 shows, for $n = 50, 100, \ldots, 350$, the average value of $\Psi^{(1)}$ after optimization relative to $\Psi^{(1)}$ and relative to $\Psi^{(2)}, \Psi^{(1)}$, the average value of $\Psi^{(2)}$ after optimization relative to $\Psi^{(2)}$ and relative to $\Psi^{(1)}, \Psi^{(2)}$, and the average value of $\Psi^{(3)}$ after optimization relative to $\Psi^{(3)}$, relative to $\Psi^{(1)}, \Psi^{(3)}$, and relative to $\Psi^{(2)}, \Psi^{(3)}$ for 5-clustering of the sequence. The length of the longest cluster increased on average when comparing the optimization relative to $\Psi^{(3)}$ with the optimization relative to $\Psi^{(1)}, \Psi^{(3)}$, and then with the optimization relative to $\Psi^{(2)}, \Psi^{(3)}$.

We also considered bi-criteria optimization of k-clusterings. Table 11.3 shows the minimum, maximum, and the average number of Pareto optimal points for bi-criteria optimization problems relative to $\Psi^{(1)}$ and $\Psi^{(2)}$ for $n = 50, 100, \ldots, 250$ and $k = 5, 10, 15$. The maximum number of Pareto optimal points was 20 among all tested scenarios. Table 11.4 shows the number of cases with totally optimal k-clusterings relative to $\Psi^{(1)}$ and $\Psi^{(2)}$. For 15-clustering, there were at least 6 cases with totally optimal clusterings among all testes scenarios. The set of Pareto optimal

Table 11.3 Number of Pareto optimal points (in format $_{min}average_{max}$ among 100 trials) for bi-criteria optimization problem relative to $\Psi^{(1)}$ and $\Psi^{(2)}$, numbers from [0, 100]

	k		
n	5	10	15
50	$_{1}2.02_{6}$	$_{1}2.16_{7}$	$_{1}2.09_{6}$
100	$_{1}2.89_{13}$	$_{1}2.95_{13}$	$_{1}2.88_{8}$
150	$_{1}4.09_{13}$	$_{1}3.82_{11}$	$_{1}3.03_{7}$
200	$_{1}4.72_{20}$	$_{1}3.74_{10}$	$_{1}3.56_{10}$
250	$_{1}4.72_{19}$	$_{1}3.56_{12}$	$_{1}3.79_{9}$

Table 11.4 Number of cases with totally optimal k-clusterings relative to $\Psi^{(1)}$ and $\Psi^{(2)}$ among 100 trials, numbers from [0, 100]

	k		
n	5	10	15
50	49	33	34
100	27	22	27
150	18	11	16
200	21	11	14
250	13	12	6

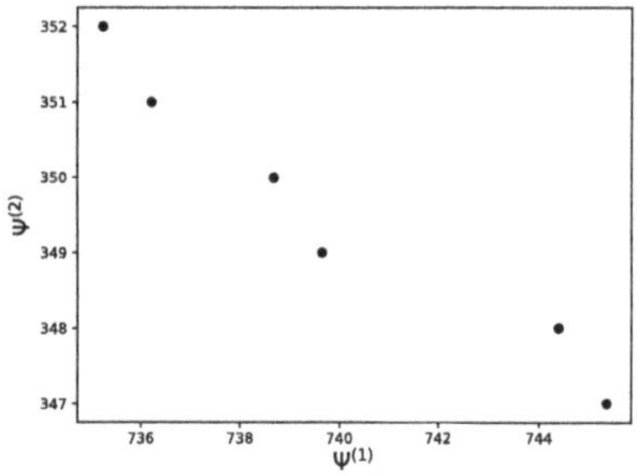

Fig. 11.4 Set of Pareto optimal points for a bi-criteria optimization of 15-clusterings relative to $\Psi^{(1)}$ and $\Psi^{(2)}$, $n = 250$

points for a bi-criteria optimization of 15-clusterings relative to $\Psi^{(1)}$ and $\Psi^{(2)}$ for $n = 250$ is depicted in Fig. 11.4.

Table 11.5 shows the minimum, maximum, and the average number of Pareto optimal points for bi-criteria optimization problems relative to $\Psi^{(2)}$ and $\Psi^{(3)}$ for $n = 50, 100, \ldots, 250$ and $k = 5, 10, 15$. The maximum number of Pareto optimal points was 10 among all tested scenarios. Table 11.6 shows the number of cases with totally optimal k-clusterings relative to $\Psi^{(2)}$ and $\Psi^{(3)}$. None of the clusterings were totally optimal relative to $\Psi^{(2)}$ and $\Psi^{(3)}$ for $n = 250$ and $k = 5, 10$. The set of Pareto optimal points for a bi-criteria optimization of 15-clusterings relative to $\Psi^{(2)}$ and $\Psi^{(3)}$ for $n = 250$ is depicted in Fig. 11.5.

Table 11.5 Number of Pareto optimal points (in format $_{min}average_{max}$ among 100 trials) for bi-criteria optimization problem relative to $\Psi^{(2)}$ and $\Psi^{(3)}$, numbers from [0, 100]

	k		
n	5	10	15
50	$_{1}3.07_{8}$	$_{1}2.29_{5}$	$_{1}1.88_{3}$
100	$_{1}3.64_{8}$	$_{1}2.58_{5}$	$_{1}2.33_{4}$
150	$_{1}4.01_{10}$	$_{1}3.04_{6}$	$_{1}2.19_{4}$
200	$_{1}3.54_{8}$	$_{1}3.24_{6}$	$_{1}2.32_{4}$
250	$_{2}3.93_{8}$	$_{2}2.98_{6}$	$_{1}2.38_{4}$

Table 11.6 Number of cases with totally optimal k-clusterings relative to $\Psi^{(2)}$ and $\Psi^{(3)}$ among 100 trials, numbers from [0, 100]

	k		
n	5	10	15
50	13	19	32
100	5	12	12
150	5	4	16
200	3	1	10
250	0	0	7

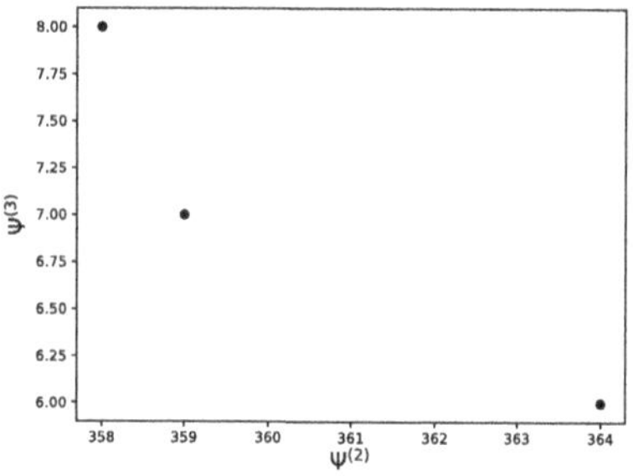

Fig. 11.5 Set of Pareto optimal points for a bi-criteria optimization of 15-clusterings relative to $\Psi^{(2)}$ and $\Psi^{(3)}$, $n = 250$

References

1. Estivill-Castro, V.: Why so many clustering algorithms: A position paper. SIGKDD Explor. Newsl. **4**(1), 65–75 (2002)
2. Grønlund, A., Larsen, K.G., Mathiasen, A., Nielsen, J.S.: Fast exact k-means, k-medians and Bregman divergence clustering in 1D. arxiv: abs/1701.07204 (2017)
3. Mankowski, M., Moshkov, M.: Extensions of dynamic programming for multi-stage combinatorial optimization. Theor. Comput. Sci. **844**, 106–132 (2020)
4. Wang, H., Song, M.: Ckmeans.1d.dp: Optimal k-means clustering in one dimension by dynamic programming. R J. **3**(2), 29–33 (2011)

Chapter 12
Optimal Bitonic Tour

The optimal bitonic tour problem is a restricted variant of the Euclidean traveling salesman problem introduced by J. L. Bentley. This problem can be solved by a dynamic programming algorithm in polynomial time [1]. A bitonic tour starts from the rightmost point, and it goes strictly right to left to the leftmost point, and then goes strictly left to right to the rightmost point. This tour passes through all given points (with the exception of the rightmost point) exactly one time. We consider a problem of finding bitonic tours with minimum length (relative to Euclidean or Manhattan distance), and a problem of finding tours, which minimize the maximum Euclidean distance between two consecutive points.

In this chapter, we design a conventional circuit without repetitions for the problem of the optimal bitonic tour, describe three cost functions for this circuit, evaluate the number of operations and the time required by the algorithms for the optimization and counting of bitonic tours, discuss an example, and show experimental results.

12.1 Definition of Circuit and Cost Functions

Let us consider a sequence $p_1, p_2, \ldots, p_n$ of n, $n \geq 2$, distinct points in two-dimensional Euclidean space. Each point p_k has corresponding nonnegative real coordinates (a_k, b_k), $k = 1, 2, \ldots, n$, such that $a_1 < a_2 < \cdots < a_n$ (points are sorted from left to right). A path is a nonempty sequence of points from the set $\{p_1, p_2, \ldots, p_n\}$. Let $1 \leq i < j \leq n$. An (i, j)-bitonic path $p_i \cdots p_1 \cdots p_j$ is a path that starts from the point p_i and goes strictly left ($a_i > \cdots > a_1$) to the point p_1 and then goes strictly right ($a_1 < \cdots < a_j$) to the point p_j. This path contains all points $p_1, p_2, \ldots, p_j$ exactly one time.

For any path α and for an arbitrary point p_k, we denote $\alpha \otimes p_k = \alpha p_k$, and for a set of paths A, we denote $A \otimes p_k = \{\alpha p_k : \alpha \in A\}$. Also, for an arbitrary path $\pi = p_l \cdots p_m$ we denote $\sim \pi = p_m \cdots p_l$ (path with the reversed order of points),

M. Mankowski and M. Moshkov, *Dynamic Programming Multi-Objective Combinatorial Optimization*, Studies in Systems, Decision and Control 331,
https://doi.org/10.1007/978-3-030-63920-4_12

and for the set A, we denote $\sim A = \{\sim \alpha : \alpha \in A\}$. We denote by $\Pi(i, j)$ the set of all (i, j)-bitonic paths. We have $\Pi(1, 2) = \{p_1 p_2\}$ and, for $1 \le i < j \le n$, $j \ne 2$,

$$\Pi(i, j) = \Pi(i, j-1) \otimes p_j, \text{ if } i < j-1 ,$$

$$\Pi(i, j) = \bigcup_{1 \le k < j-1} ((\sim \Pi(k, j-1)) \otimes p_j), \text{ if } i = j-1 .$$

The set of all bitonic tours under consideration is defined by $\Pi(n, n) = (\sim \Pi(n-1, n)) \otimes p_n$. Among two bitonic tours π and $\sim \pi$, the set $\Pi(n, n)$ contains the only one in which two last points are p_{n-1} and p_n. We will consider bitonic tours as a special case of bitonic paths starting and ending in p_n.

The number n is the dimension of the problems under consideration related to the optimization of bitonic tours. To describe an instance of a problem, we need to specify coordinates of n points $p_1, \ldots, p_n$, i.e., $2n$ real numbers. Note that n is at most the number of numeric parameters of the problem.

Let us assume that the coordinates of the points $p_1, \ldots, p_n$ are integers. The binary representations of these $2n$ coordinates form the representation of the problem instance (we will omit the word instance). It is clear that n is at most the length of the problem representation.

Set $U = \Pi(n, n) \cup \bigcup_{1 \le i < j \le n} \Pi(i, j)$. We now describe a circuit S without repetitions over the set U. For $1 \le i < j \le n$ and $i < j-1$, we define a concatenation function $\mu_{i,j-1,j}$, from $\Pi(i, j-1)$ to $\Pi(i, j)$, such that $\mu_{i,j-1,j} = x \otimes p_j$. For $1 \le i < j \le n$, $i = j-1$, and $1 \le k < j-1$, we define a concatenation function $\mu_{i,k,j}$ from $\Pi(k, j-1)$ to $\Pi(i, j)$, such that $\mu_{i,k,j} = (\sim x) \otimes p_j$. There is also a concatenation function $\mu_{n,n-1,n}$, from $\Pi(n-1, n)$ to $\Pi(n, n)$ such that $\mu_{n,n-1,n} = (\sim x) \otimes p_n$.

The circuit S contains one input node $v(1, 2)$ labeled with $\{p_1 p_2\}$. Let $1 \le i < j \le n$. If $i < j-1$ and $j < n$, the circuit S contains a unifying node $v(i, j)$ labeled with $\cup$ and a functional node $v(i, j-1, j)$ labeled with $\mu_{i,j-1,j}$. If $i = j-1$, $1 \le k < j-1$, and $j \ne 2$, the circuit S contains a unifying node $v(i, j)$ labeled with $\cup$ and a functional node $v(i, k, j)$ labeled with $\mu_{i,k,j}$. The circuit S also contains one unifying node $v(n, n)$ labeled with $\cup$ and $*$, which is considered as the output, and one functional node $v(n, n-1, n)$ labeled with the function $\mu_{n,n-1,n}$.

Let $1 \le i < j \le n$. If $i < j-1$ and $j < n$, the unifying node $v(i, j)$ has one entering edge, which leaves the node $v(i, j-1, j)$, and the functional node $v(i, j-1, j)$ has one entering edge, which leaves the node $v(i, j-1)$. If $i = j-1$, $1 \le k < j-1$, and $j \ne 2$, the unifying node $v(i, j)$ has $j-2$ entering edges, which leave the nodes $v(i, k, j)$, $k = 1, \ldots, j-2$, and the functional node $v(i, k, j)$ has one entering edge, which leaves the node $v(k, j-1)$. Also, the unifying node $v(n, n)$ has one entering edge, which leaves the node $v(n, n-1, n)$, and the functional node $v(n, n-1, n)$ has one entering edge, which leaves the node $v(n-1, n)$. We will not label edges, which enter functional nodes, with the number 1.

Let $1 \le i < j \le n$. One can show that for any node $v(i, j)$, $S(v(i, j)) = \Pi(i, j)$. If $i < j-1$ and $j < n$, then $S(v(i, j-1, j)) = \Pi(i, j-1) \otimes p_j$. If $i = j-1$

and $1 \le k < j - 1$, then $S(v(i,k,j)) = (\sim \Pi(k, j-1)) \otimes p_j$. Also, $S(v(n,n)) = \Pi(n,n)$ and $S(v(n, n-1, n)) = (\sim \Pi(n-1, n)) \otimes p_n$.

The described circuit will be considered as the preliminary version of the circuit S. The final version of S is obtained from the preliminary version of S by the removal of all nodes such that there is no directed path from the considered node to the output. We also remove all edges that are incident to the removed nodes. Later, S without mentioning the version denotes the final version of the circuit S.

The preliminary version of the circuit S contains one input node, at most n unifying nodes with at most n entering edges, at most n^2 unifying nodes with one entering edge, and at most $2n^2$ functional nodes with one entering edge. Hence, this circuit has at most $1 + n + 3n^2$ nodes and at most $4n^2$ edges.

We now prove that S is a circuit without repetitions. It is clear that all considered concatenation functions $\mu_{i,j-1,j}$, $\mu_{i,k,j}$, and $\mu_{n,n-1,n}$ are injective. Let v_1 and v_2 be different nodes from $V_i(S) \cup V_f(S)$. We now show that $S(v_1) \cap S(v_2) = \emptyset$. Let v_1 be the input node and v_2 be a functional node. It is clear that sets $S(v_1)$ and $S(v_2)$ do not contain common elements since $S(v_1)$ contains the only bitonic path $p_1 p_2$ but $S(v_2)$ does not contain any path starting at p_1 and ending in p_2. Let v_1 and v_2 be functional nodes and $v_1 = v(i_1, k_1, j_1)$, $v_2 = v(i_2, k_2, j_2)$. If $i_1 \ne i_2$, then $S(v(i_1, k_1, j_1))$ contains only bitonic paths starting at p_{i_1}, where $S(v(i_2, k_2, j_2))$ contains only bitonic paths starting at p_{i_2}, thus $S(v_1) \cap S(v_2) = \emptyset$. Similarly, if $j_1 \ne j_2$, then $S(v(i_1, k_1, j_1))$ contains only bitonic paths ending at p_{j_1}, where $S(v(i_2, k_2, j_2))$ contains only bitonic paths ending at p_{j_2}, thus $S(v_1) \cap S(v_2) = \emptyset$. If $k_1 \ne k_2$, then $S(v(i_1, k_1, j_1))$ contains only bitonic paths with a penultimate point p_{k_1}, where $S(v(i_2, k_2, j_2))$ contains only bitonic paths with a penultimate point p_{k_2}. Therefore, $S(v_1) \cap S(v_2) = \emptyset$.

We now define a cost function $\Psi^{(t)}$, $t = 1, 2, 3$, for the circuit S given by a function $\psi^{(t)}$ and function $\psi_v^{(t)}$ for each $v \in V_f(S)$. Denote

$$B = \{x + y, x - y, x \cdot y, \max(x, y), |x|, \sqrt{x}\} .$$

One of the notions of optimality is to find the shortest bitonic path. For arbitrary points p_i and p_j with coordinates (a_i, b_i) and (a_j, b_j), we define Euclidean distance by $d(p_i, p_j) = \sqrt{(a_i - a_j)^2 + (b_i - b_j)^2}$ and Manhattan distance by $d_1(p_i, p_j) = |a_i - a_j| + |b_i - b_j|$. The cost function $\Psi^{(1)}$ is defined by the functions $\psi^{(1)}(p_1 p_2) = d(p_1, p_2)$ and $\psi^{(1)}_{v(i,k,j)}(x) = x + d(p_k, p_j)$, $v(i,k,j) \in V_f(S)$. Similarly, $\Psi^{(2)}$ is defined by the functions $\psi^{(2)}(p_1 p_2) = d_1(p_1, p_2)$ and $\psi^{(2)}_{v(i,k,j)}(x) = x + d_1(p_k, p_j)$, $v(i,k,j) \in V_f(S)$. The cost functions $\Psi^{(1)}$ and $\Psi^{(2)}$ are strictly increasing. To compute $\psi^{(1)}_{v(i,k,j)}$ we need seven operations from the set B while to compute $\psi^{(1)}$ we need six operations from B. To compute $\psi^{(2)}_{v(i,k,j)}$ we need six operations from B while to compute $\psi^{(2)}$ we need five operations from B. If all coordinates of the points $p_1, \dots, p_n$ are integers, then $\Psi^{(2)}$ is a strictly increasing nonnegative integer cost function for S.

We also consider a variant of the bottleneck bitonic tour problem, where we aim to minimize the longest Euclidean distance between two consecutive points on the bitonic path. The cost function $\Psi^{(3)}$ is defined by the functions $\psi^{(3)}(p_1 p_2) = d(p_1, p_2)$ and $\psi^{(3)}_{v(i,k,j)}(x) = \max(x, d(p_k, p_j))$, $v(i, k, j) \in V_f(S)$. The cost function $\Psi^{(3)}$ is increasing. To compute $\psi^{(3)}_{v(i,k,j)}$, we need seven operations from B while to compute $\psi^{(3)}$ we need six operations from B.

All these cost functions are adapted (see Sect. 2.4). They are described by formulas over the set of elementary operations B.

12.2 Time Complexity of Algorithms

The algorithms $\mathscr{A}_1$, $\mathscr{A}_2$, and $\mathscr{A}_4$ work with the scheme of the circuit S (see Sect. 2.4). The preliminary version of the circuit S is completely described. Therefore, the time complexity of the scheme of the preliminary version of S construction is linear depending on the number of its nodes and edges, i.e., $O(n^2)$. Using breadth-first search from the output node in the scheme of the preliminary version of S with the reverse direction of the edges, we can find in the scheme of the preliminary version of S all nodes such that there is no directed path from the considered node to the output and construct the scheme of S in $O(n^2)$ time. In total, we need $O(n^2)$ time to construct the scheme of S.

Note that the algorithms $\mathscr{A}_1$ and $\mathscr{A}_2$ can also work with the schemes of edge-preserving subcircuits of S obtained as a result of the algorithm $\mathscr{A}_1$ work—see Remark 3.1.

Denote $B^+ = B \cup \{x : y\}$, where $x : y$ is the operation of comparison with values $x < y$, $x = y$, and $x > y$.

Let the coordinates of all points $p_1, \ldots, p_n$ be integers. Denote

$$\rho = a_n + \max\{b_1, \ldots, b_n\} .$$

One can show that $N_{\Psi^{(2)}}(S) \leq \rho n$, where $N_{\Psi^{(2)}}(S) = \max\{|\Psi^{(2)}(\delta)| : \delta \in U(S)\}$ and $U(S) = \bigcup_{v \in V(S)} S(v)$. One can also show that $\log_2 \rho$ is at most the length of the problem representation.

We can use the algorithm $\mathscr{A}_1$ for multi-stage optimization of bitonic tours. Using Theorem 3.1, we evaluate the number of operations required by the algorithm $\mathscr{A}_1$ working with the scheme of an edge-preserving subcircuit T of the circuit S and a cost function $\Psi^{(t)}$ for S, where $t \in \{1, 2, 3\}$.

The scheme of the circuit T contains one input node. We need at most six operations from B to process this node.

The scheme of T contains at most n unifying nodes with at most n entering edges. To treat these nodes, the algorithm $\mathscr{A}_1$ makes at most $(2n - 1)n \leq 2n^2$ comparisons. Also, the scheme contains at most n^2 unifying nodes with one entering edge. To treat these nodes the algorithm $\mathscr{A}_1$ makes at most n^2 comparisons.

The scheme of T contains at most $2n^2$ functional nodes. For each functional node v, the algorithm makes at most seven operations from B to compute $\psi_v^{(t)}$. The number of operations from B for all functional nodes is at most $14n^2$. The total number of operations from the set B^+ made by $\mathcal{A}_1$ is at most $17n^2 + 6 \leq 19n^2$, i.e., it is polynomial depending on the dimension n.

In the software model of computation, for the considered scheme of the circuit and cost function, the algorithm $\mathcal{A}_1$ has polynomial time complexity $O(n^2)$ depending on the problem dimension n.

Let $t = 2$ and coordinates of all points $p_1, \ldots, p_n$ be integers. Using Remark 3.2 we obtain that, for the algorithm $\mathcal{A}_1$, the number ρn is an upper bound on the absolute values of the numbers at the inputs of the operations $\psi_v^{(2)}$ related to the functional nodes and at the inputs of comparisons related to the unifying nodes of S. It is easy to check that, for the algorithm $\mathcal{A}_1$, the number ρn is an upper bound on the absolute values of the numbers at the inputs of operations from B in formulas for the functions $\psi_v^{(2)}$ and for the function $\psi^{(2)}$. The algorithm $\mathcal{A}_1$ uses only the operations $x + y$, $x - y$, $|x|$, and comparisons. Therefore, in the integer model of computation, for the considered scheme of the circuit and the cost function $\Psi^{(2)}$, the algorithm $\mathcal{A}_1$ has the time complexity $O(n^2 \log(\rho n))$, which is polynomial depending on the length of the problem representation (n and $\log_2 \rho$ are at most the length of the problem representation).

To count all possible bitonic tours described by the circuit T, we apply the algorithm $\mathcal{A}_2$. We use Theorem 3.2 to evaluate the number of operations required by $\mathcal{A}_2$. The scheme of T contains at most n unifying nodes with at most n entering edges each. Therefore, we need at most n^2 additions to treat these unifying nodes. For each of at most $2n^2$ functional and at most n^2 unifying nodes with one entering edge, we do not need any operations. To treat the input node, we do not need any operations. Thus, the total number of operations from B^+ (additions) made by $\mathcal{A}_2$ is at most n^2, i.e., it is polynomial depending on the dimension n.

In the software model of computation, for the considered scheme of the circuit, the algorithm $\mathcal{A}_2$ has polynomial time complexity $O(n^2)$ depending on the problem dimension n.

We now evaluate the number $|S(v(n, n))|$, which is equal to the number of bitonic tours in the set $\Pi(n, n)$. One can show that it is at most 2^{n-2}: to specify a bitonic tour, it is enough to describe a set of points in the tour between p_n and p_1, i.e., to describe a subset of the set $\{p_2, \ldots, p_{n-1}\}$. It is clear that the circuit S is reachable. Using Remark 3.4, we obtain that in the integer model of computation, for the considered scheme of the circuit, the algorithm $\mathcal{A}_2$ has the time complexity $O(n^3)$, which is polynomial depending on the length of the problem representation.

Let us consider two distinct cost functions $\Theta^{(1)}, \Theta^{(2)} \in \{\Psi^{(t)} : t = 1, 2, 3\}$. We can use the algorithm $\mathcal{A}_4$ to find the set of Pareto optimal points for the problem of bi-criteria optimization of bitonic tours relative to $\Theta^{(1)}$ and $\Theta^{(2)}$.

We denote

$$M = M_{\Theta^{(1)},\Theta^{(2)}}(S) = \max\{|Par(t_{\Theta^{(1)},\Theta^{(2)}}(S, v))| : v \in V(S)\} .$$

To evaluate the number of operations of the algorithm $\mathcal{A}_4$, we use Theorem 4.1. To treat the input node of the scheme of S, we need at most 12 operations from B.

The scheme of S contains at most $2n^2$ functional nodes with one entering edge each. To process one functional node, the algorithm $\mathcal{A}_4$ makes at most $14M$ operations from B and at most $4M \log_2 M$ comparisons. Therefore, the algorithm needs at most $36n^2M \log_2 M$ operations from B^+ to process all functional nodes.

For all at most n unifying nodes with at most n entering edges each, the algorithm $\mathcal{A}_4$ makes at most $4n^2M \log_2(nM)$ comparisons. For all at most n^2 unifying nodes with one entering edge each, the algorithm $\mathcal{A}_4$ makes at most $4n^2M \log_2 M$ comparisons. Then, to process all unifying nodes, the algorithm needs at most $8n^2M \log_2(nM)$ comparisons.

Hence, the algorithm $\mathcal{A}_4$ requires at most $44n^2M \log_2(nM)$ operations from B^+.

Let the coordinates of all points $p_1, \ldots, p_n$ be integers and $\Theta^{(1)} = \Psi^{(2)}$. By Proposition 4.4, $M \leq \rho n + 1$. Then, the algorithm $\mathcal{A}_4$ requires at most $44n^2(\rho n + 1) \log_2(n(\rho n + 1))$ operations from the set B^+, i.e., pseudo-polynomial number of operations (polynomial depending on the dimension n and the numeric parameter ρ). In the software model of computation, for the considered scheme of the circuit and pair of cost functions, the algorithm $\mathcal{A}_4$ has the time complexity $O(n^3\rho \log(n\rho))$, which is pseudo-polynomial (polynomial depending on the dimension n and the numeric parameter ρ).

Let $\Theta^{(1)} = \Psi^{(3)}$. Then $|\{\Psi^{(3)}(\delta) : \delta \in U(S)\}| \leq n^2$ since the number of different pairs of points is at most n^2. By Proposition 4.5, $M_{\Theta^{(1)},\Theta^{(2)}}(S) \leq n^2$. Since $M \leq n^2$, the algorithm $\mathcal{A}_4$ makes at most $132n^4 \log_2 n$ operations from B^+, i.e., polynomial number of operations depending on the dimension n. In the software model of computation, for the considered scheme of the circuit and pair of cost functions, the algorithm $\mathcal{A}_4$ has polynomial time complexity $O(n^4 \log n)$ depending on the dimension n.

When coordinates of the points $p_1, \ldots, p_n$ are integers, we have only one integer cost function $\Psi^{(2)}$. Therefore, we cannot consider the integer model of computation in the case of the algorithm $\mathcal{A}_4$.

12.3 Example

We consider five points $p_1 = (0, 1)$, $p_2 = (1, 0)$, $p_3 = (2, 0)$, $p_4 = (3, 2)$, $p_5 = (4, 1)$ in the Euclidean plane—see Fig. 12.1.

We construct the circuit S_{OBT}, which describes all bitonic tours for the considered points that are finished by p_4p_5 (Fig. 12.2). We apply to the circuit S_{OBT} the procedure of optimization relative to $\Psi^{(2)}$. As a result, we have the circuit $S_{OBT}^{\Psi^{(2)}}$, which can be obtained from the circuit S_{OBT} by the removal of edges labeled with the number 1. The circuit $S_{OBT}^{\Psi^{(2)}}$ describes the two bitonic tours: $p_5p_1p_2p_3p_4p_5$ and $p_5p_3p_2p_1p_4p_5$. We apply to the circuit $S_{OBT}^{\Psi^{(2)}}$ the procedure of optimization relative to $\Psi^{(1)}$. As a result, we have the circuit $S_{OBT}^{\Psi^{(2)},\Psi^{(1)}}$, which can be obtained from the circuit S_{OBT} by

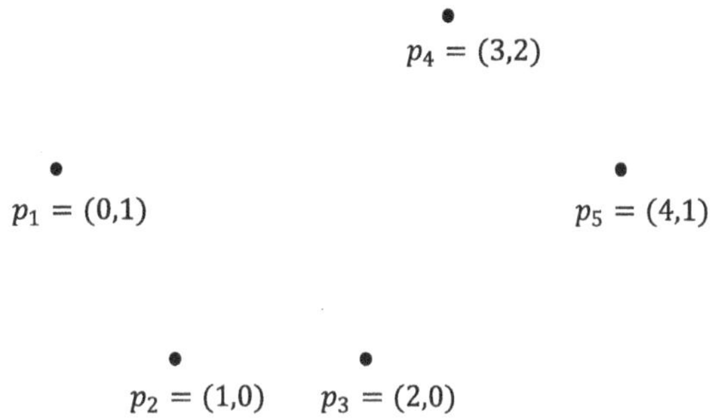

Fig. 12.1 Points $p_1, \ldots, p_5$

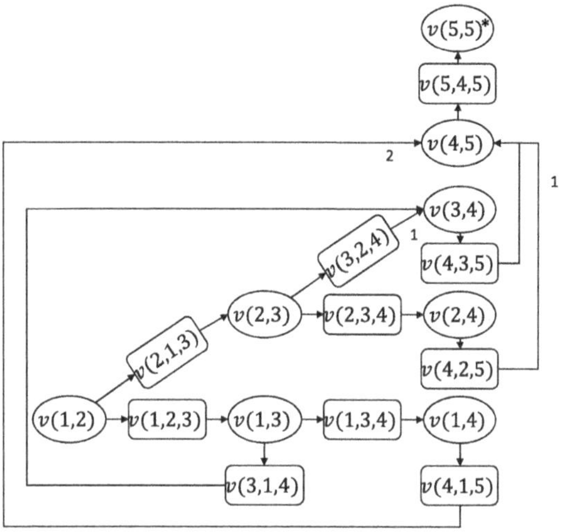

Fig. 12.2 Circuit S_{OBT} for the sequence $p_1, \ldots, p_5$

the removal of edges labeled with the numbers 1 and 2. The circuit $S_{OBT}^{\Psi^{(2)},\Psi^{(1)}}$ describes the only one tour $p_5 p_3 p_2 p_1 p_4 p_5$.

We also apply the procedure of bi-criteria optimization relative to $\Psi^{(2)}$ and $\Psi^{(1)}$ to the circuit S_{OBT}. The set of Pareto optimal points for each node of S_{OBT} is depicted in Fig. 12.3.

12.4 Experiments

We experimentally tested various scenarios for optimization of bitonic tours. Each experiment was repeated 100 times.

We randomly generated n points $p_1, p_2, \ldots, p_n$ with nonnegative integer coordinates $(x_i, y_i), i = 1, 2, \ldots, n$, where $x_i \in [0, x_{max}]$ and $y_i \in [0, y_{max}]$.

Table 12.1 shows for $n = 50, 100, \ldots, 300$ and $x_{max} = y_{max} = 1000$, the average number of bitonic tours described by the circuit before optimization, after optimiza-

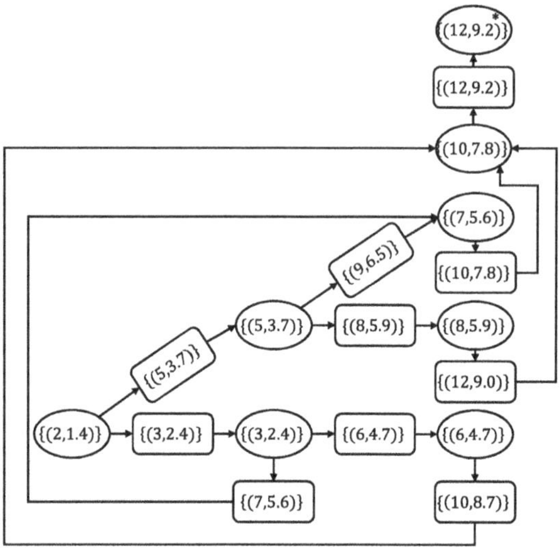

Fig. 12.3 Sets of Pareto optimal points for all nodes of the circuit S_{OBT}

Table 12.1 Average number of bitonic tours described by the circuit before optimization, after optimization relative to $\Psi^{(1)}$, relative to $\Psi^{(2)}$, and relative to $\Psi^{(3)}$ with $x_{max} = y_{max} = 1000$ among 100 trials

n	$\emptyset$	$\Psi^{(1)}$	$\Psi^{(2)}$	$\Psi^{(3)}$
50	1.41×10^{14}	1	1.67	2.97×10^{12}
100	1.58×10^{29}	1	2.24	1.04×10^{27}
150	1.78×10^{44}	1	2.82	1.34×10^{42}
200	2.01×10^{59}	1	3.95	6.18×10^{55}
250	2.26×10^{74}	1	5.24	3.59×10^{69}
300	2.55×10^{89}	1	6.06	8.95×10^{84}

tion relative to $\Psi^{(1)}$, relative to $\Psi^{(2)}$, and relative to $\Psi^{(3)}$. There was always one optimal bitonic tour relative to $\Psi^{(1)}$, while there were many optimal bitonic tours relative to $\Psi^{(3)}$.

Table 12.2 shows the average value of $\Psi^{(1)}$ after optimization relative to $\Psi^{(1)}$ and relative to $\Psi^{(2)}, \Psi^{(1)}$, average value of $\Psi^{(2)}$ after optimization relative to $\Psi^{(2)}$ and relative to $\Psi^{(1)}, \Psi^{(2)}$, and average value of $\Psi^{(3)}$ after optimization relative to $\Psi^{(3)}$ and relative to $\Psi^{(1)}, \Psi^{(3)}$. For all tested n, the average value of $\Psi^{(3)}$ increases noticeably while comparing bitonic tours optimal relative to $\Psi^{(3)}$ and relative to $\Psi^{(1)}, \Psi^{(3)}$.

We also consider bi-criteria optimization of bitonic tours. Table 12.3 shows the minimum, maximum, and the average number of Pareto optimal points for bi-criteria optimization problems relative to $\Psi^{(1)}$ and $\Psi^{(2)}$ for $n = 50, 100, \ldots, 300$, $x_{max} = 1000$, and $y_{max} = 500, 1000, \ldots, 3000$. The minimum number of Pareto optimal points was one and the maximum number was nine among all tested scenarios. For all experiments, there were at least 63 (among 100) cases with totally optimal bitonic

Table 12.2 Average value of $\Psi^{(1)}$ after optimization relative to $\Psi^{(1)}$ and relative to $\Psi^{(2)}, \Psi^{(1)}$, average value of $\Psi^{(2)}$ after optimization relative to $\Psi^{(2)}$ and relative to $\Psi^{(1)}, \Psi^{(2)}$, and average value of $\Psi^{(3)}$ after optimization relative to $\Psi^{(3)}$ and relative to $\Psi^{(1)}, \Psi^{(3)}$ with $x_{max} = y_{max} = 1000$ among 100 trials

n	$\Psi^{(1)}$	$\Psi^{(2)}, \Psi^{(1)}$	$\Psi^{(2)}$	$\Psi^{(1)}, \Psi^{(2)}$	$\Psi^{(3)}$	$\Psi^{(1)}, \Psi^{(3)}$
50	8.67×10^{03}	8.68×10^{03}	9.95×10^{03}	9.96×10^{03}	583.09	604.66
100	1.60×10^{04}	1.60×10^{04}	1.76×10^{04}	1.76×10^{04}	562.61	593.68
150	2.34×10^{04}	2.34×10^{04}	2.50×10^{04}	2.50×10^{04}	603.31	640.06
200	3.09×10^{04}	3.09×10^{04}	3.26×10^{04}	3.26×10^{04}	583.10	638.45
250	3.83×10^{04}	3.83×10^{04}	4.00×10^{04}	4.00×10^{04}	580.38	633.78
300	4.56×10^{04}	4.56×10^{04}	4.74×10^{04}	4.74×10^{04}	580.88	637.93

Table 12.3 Number of Pareto optimal points (in format $_{min}average_{max}$ among 100 trials) for bi-criteria optimization problem relative to $\Psi^{(1)}, \Psi^{(2)}$, $x_{max} = 1000$

	y_{max}					
n	500	1000	1500	2000	2500	3000
50	$_{1}1.46_{5}$	$_{1}1.37_{9}$	$_{1}1.36_{5}$	$_{1}1.23_{3}$	$_{1}1.34_{4}$	$_{1}1.44_{4}$
100	$_{1}1.35_{6}$	$_{1}1.28_{3}$	$_{1}1.34_{4}$	$_{1}1.33_{4}$	$_{1}1.22_{3}$	$_{1}1.40_{8}$
150	$_{1}1.23_{3}$	$_{1}1.33_{6}$	$_{1}1.37_{4}$	$_{1}1.30_{4}$	$_{1}1.35_{7}$	$_{1}1.34_{4}$
200	$_{1}1.20_{4}$	$_{1}1.31_{4}$	$_{1}1.23_{2}$	$_{1}1.14_{2}$	$_{1}1.26_{4}$	$_{1}1.19_{2}$
250	$_{1}1.21_{6}$	$_{1}1.21_{2}$	$_{1}1.15_{4}$	$_{1}1.19_{4}$	$_{1}1.17_{3}$	$_{1}1.18_{2}$
300	$_{1}1.14_{4}$	$_{1}1.18_{2}$	$_{1}1.20_{4}$	$_{1}1.11_{2}$	$_{1}1.23_{3}$	$_{1}1.24_{3}$

Table 12.4 Number of cases with totally optimal bitonic tours relative to $\Psi^{(1)}$ and $\Psi^{(2)}$ among 100 trials, $x_{max} = 1000$

	y_{max}					
n	500	1000	1500	2000	2500	3000
50	67	76	73	80	69	63
100	75	74	72	72	80	74
150	79	77	71	77	72	71
200	84	76	77	86	79	81
250	86	79	87	83	84	82
300	88	82	82	89	78	77

tours relative to $\Psi^{(1)}$ and $\Psi^{(2)}$—see Table 12.4. The set of Pareto optimal points for a bi-criteria optimization of bitonic tours ($n = 300$, $x_{max} = 1000$, $y_{max} = 3000$) relative to $\Psi^{(1)}$ and $\Psi^{(2)}$ is depicted in Fig. 12.4.

Table 12.5 shows the minimum, maximum, and the average number of Pareto optimal points for bi-criteria optimization problems relative to $\Psi^{(1)}$ and $\Psi^{(3)}$ for $n = 50, 100, \ldots, 300$, $x_{max} = 1000$, and $y_{max} = 500, 1000, \ldots, 3000$. The mini-

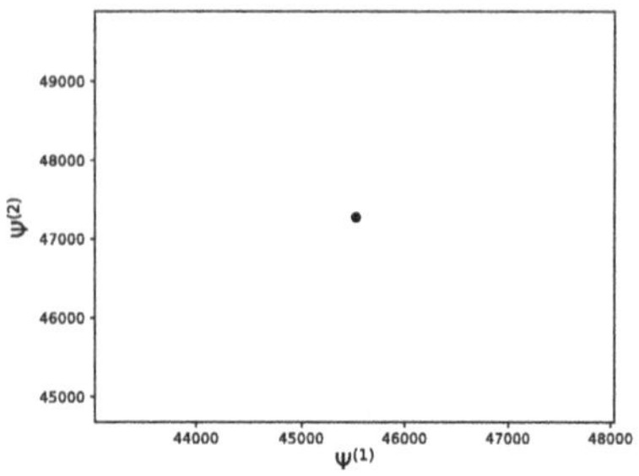

Fig. 12.4 Set of Pareto optimal points for a bi-criteria optimization of bitonic tours ($n = 300$, $x_{max} = 1000$ and $y_{max} = 3000$) relative to $\Psi^{(1)}$ and $\Psi^{(2)}$

Table 12.5 Number of Pareto optimal points (in format $_{min}average_{max}$ among 100 trials) for bi-criteria optimization problem relative to $\Psi^{(1)}$, $\Psi^{(3)}$, $x_{max} = 1000$

	y_{max}					
n	500	1000	1500	2000	2500	3000
50	$_{1}1.42_{6}$	$_{1}1.49_{5}$	$_{1}1.58_{5}$	$_{1}1.42_{5}$	$_{1}1.54_{5}$	$_{1}1.47_{4}$
100	$_{1}2.00_{9}$	$_{1}1.75_{8}$	$_{1}1.82_{6}$	$_{1}2.09_{8}$	$_{1}2.02_{6}$	$_{1}1.91_{8}$
150	$_{1}2.33_{7}$	$_{1}2.38_{8}$	$_{1}2.38_{9}$	$_{1}2.56_{9}$	$_{1}2.55_{9}$	$_{1}2.62_{8}$
200	$_{1}2.68_{11}$	$_{1}2.74_{10}$	$_{1}2.83_{11}$	$_{1}2.86_{10}$	$_{1}2.72_{13}$	$_{1}3.04_{9}$
250	$_{1}3.22_{11}$	$_{1}3.25_{11}$	$_{1}2.90_{10}$	$_{1}3.42_{12}$	$_{1}3.14_{10}$	$_{1}3.32_{12}$
300	$_{1}3.78_{12}$	$_{1}3.36_{14}$	$_{1}2.89_{11}$	$_{1}3.38_{12}$	$_{1}3.50_{16}$	$_{1}2.92_{12}$

Table 12.6 Number of cases with totally optimal bitonic tours relative to $\Psi^{(1)}$ and $\Psi^{(3)}$ among 100 trials, $x_{max} = 1000$

	y_{max}					
n	500	1000	1500	2000	2500	3000
50	74	71	64	73	66	70
100	59	60	65	53	54	54
150	53	53	43	46	49	44
200	52	44	46	40	42	44
250	43	43	48	35	44	42
300	45	40	53	43	40	48

mum number of Pareto optimal points was one and the maximum number was 16 among all tested scenarios. For all experiments, there were at least 40 (among 100) cases with totally optimal bitonic tours relative to $\Psi^{(1)}$ and $\Psi^{(3)}$—see Table 12.6. The set of Pareto optimal points for a bi-criteria optimization of bitonic tours ($n = 300$, $x_{max} = 1000$, $y_{max} = 3000$) relative to $\Psi^{(1)}$ and $\Psi^{(3)}$ is depicted in Fig. 12.5.

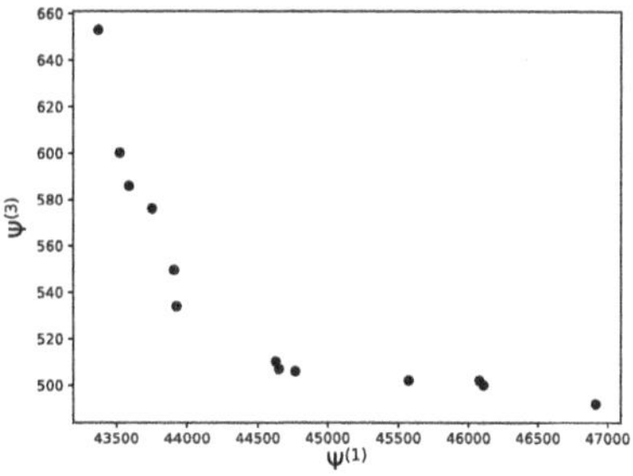

Fig. 12.5 Set of Pareto optimal points for a bi-criteria optimization of bitonic tours ($n = 300$, $x_{max} = 1000$ and $y_{max} = 3000$) relative to $\Psi^{(1)}$ and $\Psi^{(3)}$

Reference

1. Cormen, T.H., Stein, C., Rivest, R.L., Leiserson, C.E.: Introduction to Algorithms, 2nd edn. The MIT Press and McGraw-Hill Book Company (2001)

Chapter 13
Segmented Least Squares

The least squares method is widely used in statistics with common application to regression. The aim is to fit a linear function to set of points that minimizes the sum of the squares of the residuals. In some cases, fitting a linear function with a relatively small error is impossible. Keeping the linear character of approximation, the data points can be split into a sequence of segments, where to each of the segments the line given by a linear function is fitted. The optimization objectives for this problem is to minimize the total least squares error for all segments and to minimize the number of segments used. We refer to such a problem as segmented least squares. The problem and its dynamic programming solution were proposed in [2], where authors considered a single objective, which is a sum of two objectives: the total least squares error and the number of lines used multiplied by the constant weight. We consider these objectives separately. Note that different variants of the problem of points fitting by segmented straight lines along with dynamic programming solutions were considered earlier in [1].

In this chapter, we design a conventional circuit without repetitions for the problem of segmented least squares, describe two cost functions for this circuit, evaluate the number of operations and the time required by the algorithms for the optimization and counting of segmentations, discuss an example, and show experimental results.

13.1 Definition of Circuit and Cost Functions

Let us consider a sequence $p_1, p_2, \ldots, p_n$ of n, $n \geq 2$, distinct points in two-dimensional Euclidean space. Each point p_k has corresponding coordinates (x_k, y_k), $k = 1, 2, \ldots, n$, such that $x_1 < x_2 < \cdots < x_n$ (points are sorted from left to right).

M. Mankowski and M. Moshkov, *Dynamic Programming Multi-Objective Combinatorial Optimization*, Studies in Systems, Decision and Control 331,
https://doi.org/10.1007/978-3-030-63920-4_13

Let us consider segmentations of a sequence $p_1, p_2, \dots, p_n$ into nonempty, non-overlapping, and contiguous segments

$$[p_1, \dots, p_k] \cdots [p_l, \dots, p_{l+t}] \cdots [p_m, \dots, p_n] .$$

The number n is the dimension of the problems under consideration related to the optimization of segmentations. To describe an instance of a problem, we need to specify coordinates of n points $p_1, \dots, p_n$, i.e., $2n$ real numbers. Note that n is at most the number of numeric parameters of the problem.

Let us assume that the coordinates of the points $p_1, \dots, p_n$ are integers. The binary representations of these $2n$ coordinates form the representation of the problem instance (we will omit the word instance). It is clear that n is at most the length of the problem representation.

We denote by $\alpha \otimes \beta$ the concatenation $\alpha\beta$ of the sequences of segments α and β. Let λ be the empty sequence of segments. We have $\lambda \otimes \alpha = \alpha$ for any sequence α. For a set D of sequences of segments, we denote $D \otimes \alpha = \{\delta \otimes \alpha : \delta \in D\}$.

For $i = 1, \dots, n$, we denote by $T(i)$ the set of all sequences of segments for the points $p_1, \dots, p_i$. Set $T(0) = \{\lambda\}$. One can show that, for $i = 1, \dots, n$,

$$T(i) = \bigcup_{j=0,\dots,i-1} T(j) \otimes [p_{j+1}, \dots, p_i] .$$

Set $U = \bigcup_{0 \le i \le n} T(i)$. We now describe a circuit S without repetitions over the set U. We define a concatenation function $\mu_{j,i}$, $0 \le j < i \le n$, from $T(j)$ to $T(i)$ as follows: $\mu_{j,i}(x) = x \otimes [p_{j+1}, \dots, p_i]$ for any $x \in T(j)$.

The circuit S contains one input node $v(0)$ labeled with $\{\lambda\}$. For any i, $1 \le i \le n$, the circuit S contains a unifying node $v(i)$ labeled with $\cup$. For any $i, j, 0 \le j < i \le n$, the circuit S contains a functional node $v_j(i)$ labeled with the function $\mu_{j,i}$.

The unifying node $v(i)$, $i = 1, \dots, n$, has i entering edges, which leave the functional nodes $v_j(i)$ such that $0 \le j \le i - 1$. Each functional node $v_j(i)$ has one entering edge, which leaves $v(j)$. We will not label this edge with the number 1.

One can show that $S(v(i)) = T(i)$ for $i = 1, \dots, n$ and

$$S(v_j(i)) = T(j) \otimes [p_{j+1}, \dots, p_i]$$

for any functional node $v_j(i)$ of S. The node $v(n)$ is the output of S labeled with $*$.

The described circuit will be considered as the preliminary version of the circuit S. The final version of S is obtained from the preliminary version of S by the removal of all nodes such that there is no directed path from the considered node to the output. We also remove all edges that are incident to the removed nodes. Later, S without mentioning the version denotes the final version of the circuit S.

The preliminary version of the circuit S contains one input node, n unifying nodes with at most n entering edges each, and at most n^2 functional nodes with one entering edge each. Hence, it has at most $n^2 + n + 1$ nodes and at most $2n^2$ edges.

We now prove that S is a circuit without repetitions. It is clear that the concatenation functions $\mu_{j,i}$ are injective. Let v_1 and v_2 be different nodes from $V_i(S) \cup V_f(S)$. We now show that $S(v_1) \cap S(v_2) = \emptyset$. Let v_1 be the input node and v_2 be a functional node. It is clear that sets $S(v_1)$ and $S(v_2)$ do not contain common elements since $S(v_1)$ contains the only empty sequence of segments but $S(v_2)$ does not contain any empty sequence of segments. Let v_1 and v_2 be functional nodes and $v_1 = v_{j_1}(i_1)$, $v_2 = v_{j_2}(i_2)$. If $i_1 \neq i_2$, then $S(v_{j_1}(i_1))$ contains only segmentations of the sequence of points $p_1, \ldots, p_{i_1}$ and $S(v_{j_2}(i_2))$ contains only segmentations of the sequence of points $p_1, \ldots, p_{i_2}$. Therefore, $S(v_1) \cap S(v_2) = \emptyset$. Let $i_1 = i_2$ and $j_1 \neq j_2$. Then the last segment in segmentations from $S(v_{j_1}(i_1))$ contains a different number of points than the last segment in segmentations from $S(v_{j_2}(i_2))$. Thus, $S(v_1) \cap S(v_2) = \emptyset$.

We now define a cost function $\Psi^{(t)}$, $t = 1, 2$, for the circuit S given by a function $\psi^{(t)}$ and function $\psi_v^{(t)}$ for each $v \in V_f(S)$. Denote

$$B = \{x + y, x - y, x \cdot y, x/y\} .$$

For a segment $[p_j, \ldots, p_i]$, denote $err([p_j, \ldots, p_i]) = \sum_{k=j}^{i} (y_k - ax_k - b)^2$, where

$$a = \frac{(i - j + 1)\sum_{k=j}^{i} x_k y_k - (\sum_{k=j}^{i} x_k)(\sum_{k=j}^{i} y_k)}{(i - j + 1)\sum_{k=j}^{i} x_k^2 - (\sum_{k=j}^{i} x_k)^2}, \quad b = \frac{\sum_{k=j}^{i} y_k - a\sum_{k=j}^{i} x_k}{j - i + 1} .$$

This is the minimum error of a line with respect to the set of points $\{p_j, \ldots, p_i\}$ [2]. If $(i - j + 1) \leq 2$, then $err([p_j, \ldots, p_i]) = 0$.

The cost function $\Psi^{(1)}$ is defined by the functions $\psi^{(1)}(\lambda) = 0$ and $\psi^{(1)}_{v_j(i)}(x) = x + err([p_j, \ldots, p_i])$, $v_j(i) \in V_f(S)$. To compute $\psi^{(1)}_{v_j(i)}$, we need at most $16n$ elementary operations from the set B. The cost function $\Psi^{(1)}$ is strictly increasing.

Besides minimizing the least squares error, we also consider the minimization of the number of segments used. The cost function $\Psi^{(2)}$ is defined by the functions $\psi^{(2)}(\lambda) = 0$ and $\psi^{(2)}_{v_j(i)}(x) = x + 1$, $v_j(i) \in V_f(S)$. To compute $\psi^{(2)}_{v_j(i)}$, we need one elementary operation from the set B. The cost function $\Psi^{(2)}$ is a strictly increasing nonnegative integer cost function.

All these cost functions are adapted (see Sect. 2.4). They are described by formulas over the set of elementary operations B.

13.2 Time Complexity of Algorithms

The algorithms $\mathcal{A}_1$, $\mathcal{A}_2$, and $\mathcal{A}_4$ work with the scheme of the circuit S (see Sect. 2.4). The preliminary version of the circuit S is completely described. Therefore, the time complexity of the scheme of the preliminary version of S construction is linear depending on the number of its nodes and edges, i.e., $O(n^2)$. Using breadth-first search from the output node in the scheme of the preliminary version of S with the reverse direction of the edges, we can find in the scheme of the preliminary version of S all nodes such that there is no directed path from the considered node to the output and construct the scheme of S in $O(n^2)$ time. In total, we need $O(n^2)$ time to construct the scheme of S.

Note that the algorithms $\mathscr{A}_1$ and $\mathscr{A}_2$ can also work with the schemes of edge-preserving subcircuits of S obtained as a result of the algorithm $\mathscr{A}_1$ work—see Remark 3.1.

Denote $B^+ = B \cup \{x : y\}$, where $x : y$ is the operation of comparison with values $x < y$, $x = y$, and $x > y$.

One can show that $N_{\Psi^{(2)}}(S) \leq n$, where $N_{\Psi^{(2)}}(S) = \max\{|\Psi^{(2)}(\delta)| : \delta \in U(S)\}$ and $U(S) = \bigcup_{v \in V(S)} S(v)$.

We can use the algorithm $\mathscr{A}_1$ for multi-stage optimization of segmentations. Using Theorem 3.1, we evaluate the number of operations from B^+ made by the algorithm $\mathscr{A}_1$ working with the scheme of an edge-preserving subcircuit T of the circuit S and a cost function $\Psi^{(t)}$ for S, where $t \in \{1, 2\}$.

The scheme of T contains one input node. We do not need any operations to process this node.

The scheme of T contains at most n unifying nodes. Each unifying node has at most n entering edges. To treat all unifying nodes, the algorithm $\mathscr{A}_1$ makes at most $(2n - 1)n \leq 2n^2$ comparisons.

The scheme of T contains at most n^2 functional nodes. For each functional node v, to compute $\psi_v^{(t)}$, the algorithm makes at most $16n$ operations from B^+ if $t = 1$ and one operation from B^+ if $t = 2$. For all functional nodes, the number of required operations is at most $16n^3$ if $t = 1$ and at most n^2 if $t = 2$. The total number of operations from B^+ made by $\mathscr{A}_1$ is at most $18n^3$ if $t = 1$ and at most $3n^2$ if $t = 2$, i.e., it is polynomial depending on the dimension n.

In the software model of computation, for the considered scheme of the circuit and cost function, the algorithm $\mathscr{A}_1$ has polynomial time complexity depending on the problem dimension n: $O(n^3)$ if $t = 1$ and $O(n^2)$ if $t = 2$.

Let $t = 2$ and coordinates of all points $p_1, \ldots, p_n$ be integers. Using Remark 3.2, we obtain that n is an upper bound on the absolute values of the numbers at the inputs of operations of the algorithm $\mathscr{A}_1$. The algorithm uses only the operations of addition and comparison. Therefore, in the integer model of computation, for the considered scheme of the circuit and the cost function $\Psi^{(2)}$, the algorithm $\mathscr{A}_1$ has the time complexity $O(n^2 \log n)$, which is polynomial depending on the length of the problem representation (n is at most the length of the problem representation).

To count all possible segmentations described by T, we apply the algorithm $\mathscr{A}_2$. We use Theorem 3.2 to evaluate the number of operations from B^+ made by $\mathscr{A}_2$. The scheme of T contains at most n unifying nodes, each with at most n entering edges. Therefore, we need at most n^2 additions to treat all unifying nodes. For each of at most n^2 functional nodes and the input node, we do not need any operations. The total number of operations from B^+ (additions) is at most n^2.

In the software model of computation, for the considered scheme of the circuit, the algorithm $\mathcal{A}_2$ has polynomial time complexity $O(n^2)$ depending on the problem dimension n.

We now evaluate the number $|S(v(n))|$, which is equal to the number of all segmentations of the sequence of points $p_1, \ldots, p_n$. One can show that this value is equal to 2^{n-1}: there is a natural one-to-one mapping of segmentations to tuples from $\{0, 1\}^n$ in which the first digit is equal to 1. For example, the tuple $(1, 1, 0, 0, 1)$ corresponds to the segmentation $[p_1, p_2][p_3, p_4][p_5]$. It is clear that the circuit S is reachable. Using Remark 3.4, we obtain that in the integer model of computation, for the considered scheme of the circuit, the algorithm $\mathcal{A}_2$ has the time complexity $O(n^3)$, which is polynomial depending on the length of the problem representation.

We can use the algorithm $\mathcal{A}_4$ to find the set of Pareto optimal points for the problem of bi-criteria optimization of segmentations relative to $\Psi^{(1)}$ and $\Psi^{(2)}$.

We denote $M = M_{\Psi^{(2)},\Psi^{(1)}}(S)$, where

$$M_{\Psi^{(2)},\Psi^{(1)}}(S) = \max\{|Par(t_{\Psi^{(2)},\Psi^{(1)}}(S, v))| : v \in V(S)\} .$$

To evaluate the number of operations from B^+ of the algorithm $\mathcal{A}_4$, we use Theorem 4.1. To treat the input node of the scheme of S, we do not need any operations.

The scheme of S contains at most n^2 functional nodes. Each node has one entering edge. To process one functional node, the algorithm $\mathcal{A}_4$ makes at most $17nM$ operations from B and at most $4M \log_2 M$ comparisons. Therefore, the algorithm needs at most $21n^3 M \log_2(2M)$ operations from B^+ to process all functional nodes.

For each of n unifying nodes with at most n entering edges, the algorithm $\mathcal{A}_4$ makes at most $4nM \log_2(nM)$ comparisons. Then, to process all unifying nodes, it needs at most $4n^2 M \log_2(nM)$ comparisons.

Hence, the algorithm $\mathcal{A}_4$ makes at most $25n^3 M \log_2(2nM)$ operations from B^+. By Proposition 4.4, $M \leq n + 1$. Then, the algorithm $\mathcal{A}_4$ makes at most

$$25n^3(n + 1)\log_2(2n(n + 1)) \leq 50(n + 1)^4 \log_2(2(n + 1))$$

operations from B^+.

In the software model of computation, for the considered scheme of the circuit and pair of cost functions, the algorithm $\mathcal{A}_4$ has polynomial time complexity $O(n^4 \log n)$ depending on the dimension n.

When coordinates of the points $p_1, \ldots, p_n$ are integers, we have only one integer cost function $\Psi^{(2)}$. Therefore, we cannot consider the integer model of computation in the case of the algorithm $\mathcal{A}_4$.

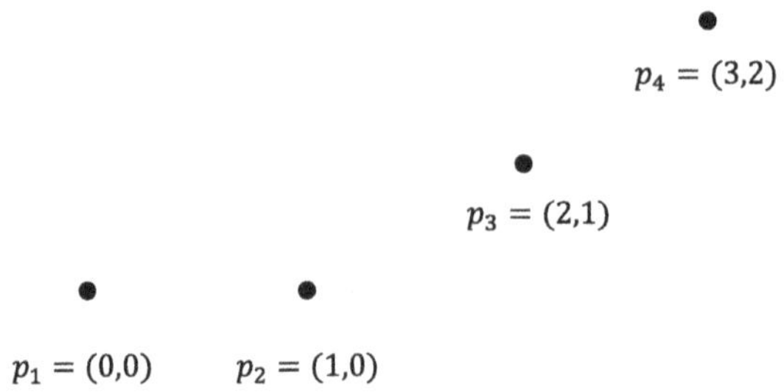

Fig. 13.1 Sequence of points p_1, p_2, p_3, p_4

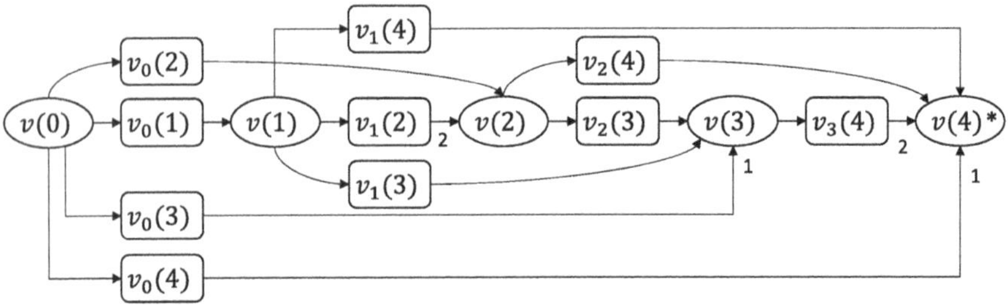

Fig. 13.2 Circuit S_{SLS} for the sequence $p_1 = (0, 0)$, $p_2 = (1, 0)$, $p_3 = (2, 1)$, $p_4 = (3, 2)$

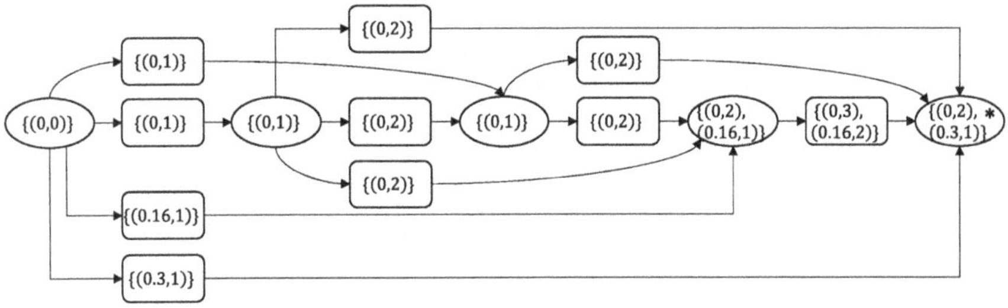

Fig. 13.3 Sets of Pareto optimal points for all nodes of the circuit S_{SLS}

13.3 Example

Let us consider the sequence of points $p_1 = (0, 0)$, $p_2 = (1, 0)$, $p_3 = (2, 1)$, $p_4 = (3, 2)$ depicted in Fig. 13.1.

We construct the circuit S_{SLS}, which describes all segmentations of this sequence (see Fig. 13.2). Apply to the circuit S_{SLS} the procedure of optimization relative to $\Psi^{(1)}$ (minimizing least squares error). As a result, we have the circuit $S_{SLS}^{\Psi^{(1)}}$, which can be obtained from the circuit S_{SLS} by the removal of edges labeled with the number 1. The circuit $S_{SLS}^{\Psi^{(1)}}$ describes six segmentations with minimum total least squares error:

$$[p_1][p_2][p_3][p_4],\ [p_1][p_2][p_3, p_4]\ ,$$
$$[p_1, p_2][p_3][p_4],\ [p_1][p_2, p_3][p_4]\ ,$$
$$[p_1, p_2][p_3, p_4],\ [p_1][p_2, p_3, p_4]\ .$$

We apply to the circuit $S_{SLS}^{\Psi^{(1)}}$ the procedure of optimization relative to $\Psi^{(2)}$. As a result, we have the circuit $S_{SLS}^{\Psi^{(1)},\Psi^{(2)}}$, which can be obtained from the circuit S_{SLS} by the removal of edges labeled with the numbers 1 and 2. The circuit $S_{SLS}^{\Psi^{(1)},\Psi^{(2)}}$ describes two segmentations $[p_1, p_2][p_3, p_4]$ and $[p_1][p_2, p_3, p_4]$ that use the minimum number of segments among segmentations with the minimum total least squares error.

We also apply to the circuit S_{SLS} the procedure of bi-criteria optimization relative to $\Psi^{(1)}$ and $\Psi^{(2)}$. The set of Pareto optimal points for each node of S_{SLS} is depicted in Fig. 13.3.

13.4 Experiments

We experimentally tested various optimization scenarios of partitioning a sequence of points into segments. Each experiment was repeated 100 times.

We randomly generated n points (x_i, y_i) in two-dimensional Euclidean space, where x_i coordinates are generated with uniform distribution from the range $[1, 7]$ and y_i is derived from the equation $y_i = 0.05x_i^2 + e_i$ or $y_i = -63 + (2144x_i)/15 - (5348x_i^2)/45 + 48x_i^3 - (91x_i^4)/9 + (16x_i^5)/15 - (2x_i^6)/45 + e_i$, where e_i is a random number from $\mathcal{N}(0, \sigma^2)$. The first equation describes the polynomial of degree

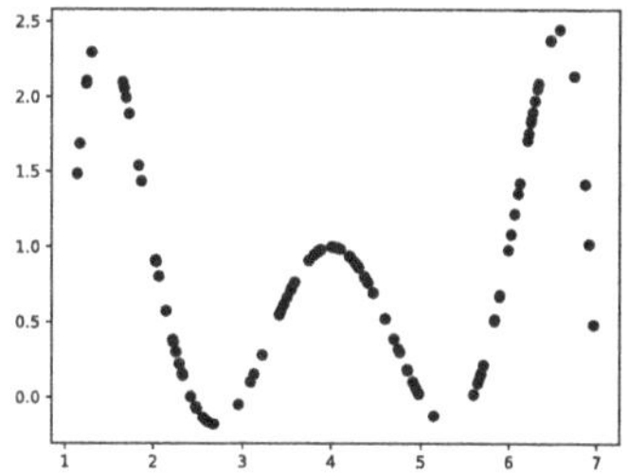

Fig. 13.4 Sequence of 100 points given by a polynomial of degree six without error noise

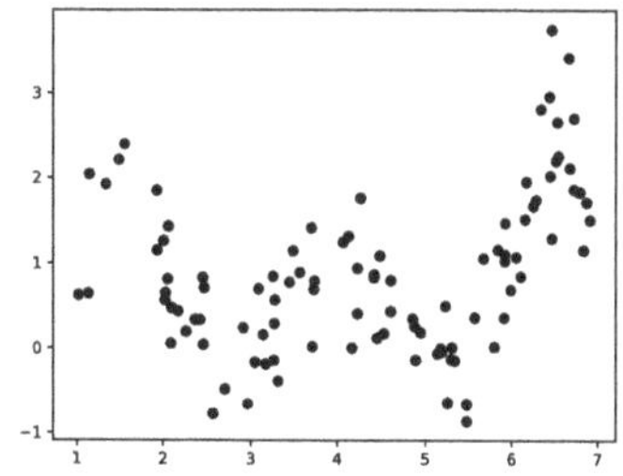

Fig. 13.5 Sequence of 100 points given by a polynomial of degree six with error noise e_i from $\mathcal{N}(0, 0.5)$

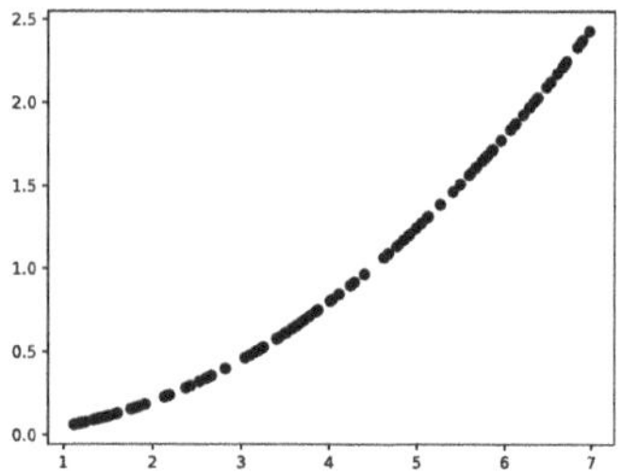

Fig. 13.6 Sequence of 100 points given by a polynomial of degree two without error noise

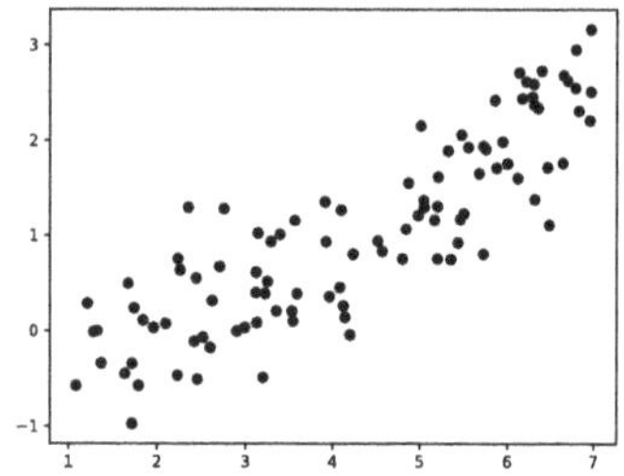

Fig. 13.7 Sequence of 100 points given by a polynomial of degree two with error noise e_i from $\mathcal{N}(0, 0.5)$

two and the second equation describes the Lagrange polynomial of degree six that passes through the points: $(1, 0)$, $(2, 1)$, $(3, 0)$, $(4, 1)$, $(5, 0)$, $(6, 1)$, $(7, 0)$. Both polynomials are noised by the error e_i derived from the normal distribution. See examples of generated points in Figs. 13.4, 13.5, 13.6, and 13.7.

Table 13.1 shows for $n = 100, 120, \ldots, 200$ points given by the polynomial of degree two, the average number of segmentations described by the circuit before optimization, after optimization relative to $\Psi^{(1)}$, and relative to $\Psi^{(2)}$. Since $\Psi^{(1)}$ (total least squares error) equals zero for each segment containing only one or two points, there are many optimal segmentations relative to $\Psi^{(1)}$. Considering optimization relative to $\Psi^{(2)}$ (number of segments) there is always one optimal solution when one segment contains all points, thus the number of such segmentations will always be one.

We also consider bi-criteria optimization relative to $\Psi^{(2)}$ and $\Psi^{(1)}$ of segmentations of points given by a polynomial of degree two and six. Regardless of the points

Table 13.1 Average number of segmentations described by the circuit before optimization, after optimization relative to $\Psi^{(1)}$, and relative to $\Psi^{(2)}$ among 100 trials

n	$\emptyset$	$\Psi^{(1)}$	$\Psi^{(2)}$
100	6.34×10^{29}	5.73×10^{20}	1
120	6.65×10^{35}	8.67×10^{24}	1
140	6.97×10^{41}	1.31×10^{29}	1
160	7.31×10^{47}	1.98×10^{33}	1
180	7.66×10^{53}	3.00×10^{37}	1
200	8.03×10^{59}	4.54×10^{41}	1

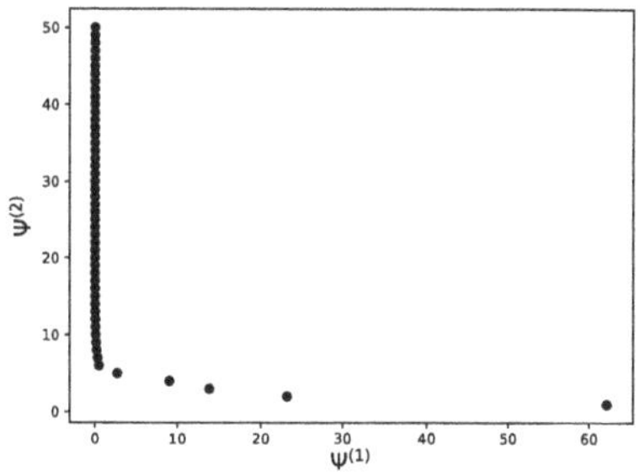

Fig. 13.8 Set of Pareto optimal points for a bi-criteria optimization of sequence of 100 points given by a polynomial of degree six (see Fig. 13.4) relative to $\Psi^{(2)}$ and $\Psi^{(1)}$

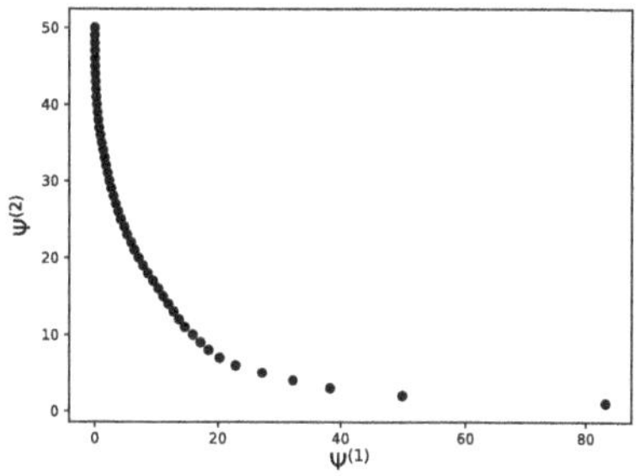

Fig. 13.9 Set of Pareto optimal points for a bi-criteria optimization of sequence of 100 points given by a polynomial of degree six with error noise e_i from $\mathcal{N}(0, 0.5)$ (see Fig. 13.5) relative to $\Psi^{(2)}$ and $\Psi^{(1)}$

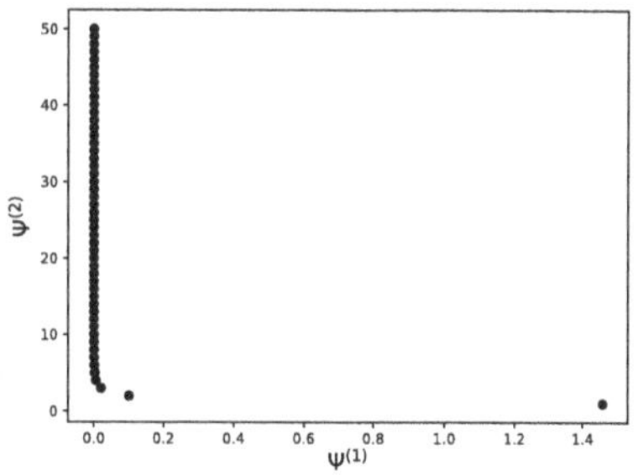

Fig. 13.10 Set of Pareto optimal points for a bi-criteria optimization of sequence of 100 points given by a polynomial of degree two (see Fig. 13.6) relative to $\Psi^{(2)}$ and $\Psi^{(1)}$

distributions, the total number of Pareto optimal points for this problem is at most (and often exactly) $\lceil n/2 \rceil$ since each segment of two points has the total least squares error equals zero. Thus, we omitted experiments analyzing the number of Pareto optimal points and number of totally optimal segmentations. Only when all the points lie on the straight line, there will be a totally optimal segmentation. We analyzed the set of POPs for points with (Figs. 13.5 and 13.7) and without (Figs. 13.4 and 13.6) noise given by a polynomial of degree two and six. In each case, our experiments ($n = 100$) resulted in 50 Pareto optimal points. For problems of points without the noise (Figs. 13.8 and 13.10) most of the Pareto optimal points had least squares error value close to zero after reaching the number of segments greater than the degree of polynomial. While using more segments, the further decrease of the error was negligible. After adding noise to the points, more segments were needed (Figs. 13.9 and 13.11) to minimize the least squares error and the Pareto optimal points are more evenly spread in the range of the $\Psi^{(1)}$.

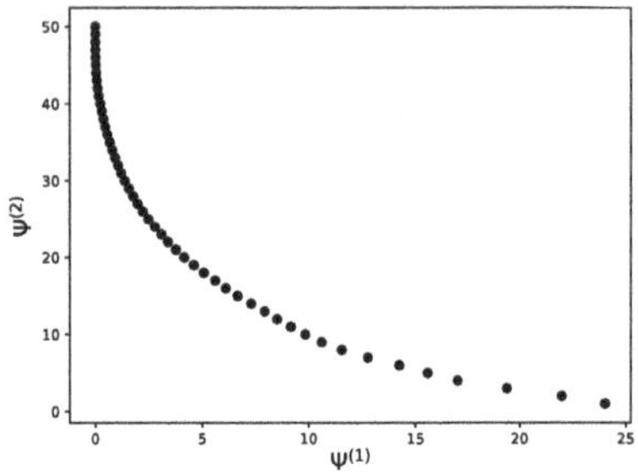

Fig. 13.11 Set of Pareto optimal points for a bi-criteria optimization of sequence of 100 points given by a polynomial of degree two with error noise e_i from $\mathcal{N}(0, 0.5)$ (see Fig. 13.7) relative to $\Psi^{(2)}$ and $\Psi^{(1)}$

References

1. Bellman, R., Roth, R.: Curve fitting by segmented straight lines. J. Am. Stat. Assoc. **64**(327), 1079–1084 (1969)
2. Kleinberg, J., Tardos, E.: Algorithm Design. Addison-Wesley Longman Publishing Co., Inc. (2005)

Part III
Multi-Stage and Bi-Criteria Optimization of Matchings in Trees

In this part, we propose an algorithm for counting matchings, and algorithms for multi-stage and bi-criteria optimization of matchings relative to different weight functions that assign positive integers (weights) to edges of the tree. We did not find a conventional circuit without repetitions that represents the set of all matchings in an arbitrary tree. Instead of a circuit, we construct a labeled forest that describes the set of matchings, and design algorithms of matching optimization and counting based on the forest. We restrict ourselves to considering only bounds on the number of operations and do not consider the time complexity of the algorithms. This part contains four chapters.

In Chap. 14, we discuss the problem of optimization of matchings in a tree and consider a labeled forest that describes the set of matchings.

In Chap. 15, we propose an algorithm for counting matchings and an algorithm for the optimization of matchings relative to a weight function.

In Chap. 16, we describe an additional tool for the study of Pareto optimal points (fusion of sets of POPs), propose an algorithm for bi-criteria optimization of matchings relative to two weight functions, and discuss the notion of a totally optimal matching.

In Chap. 17, we consider an example of a tree and its matching optimization, and discuss results of computer experiments with randomly generated trees and weight functions.

Chapter 14
Discussion of Matching Optimization Problem and Representation of Matchings

In this chapter, we discuss the problem of optimization of matchings in trees including its relation to the kidney paired donation. We did not find a conventional circuit without repetitions that represents the set of all matchings for an arbitrary tree. Instead of a circuit, we construct a labeled forest to describe the set of matchings [7]. We use this forest in multi-stage and bi-criteria optimization and counting matchings.

14.1 Discussion of Matching Optimization Problem

Let G be an undirected graph in which edges and nodes have positive weights. A matching in G is a set of edges without common nodes. We consider three optimization problems related to matchings: (c) maximization of the cardinality of matching, (n) maximization of the sum of weights of nodes in matching, and (e) maximization of the sum of weights of edges in matching. The considered problems can be solved in polynomial time [1, 5]. A linear time algorithm for the maximization of the cardinality of a matching in a tree can be found in [9].

It is known [2, 4, 8] that each solution to the problem (n) is also a solution to the problem (c). The situation with the problem (e) is different: it is possible that a solution to the problem (e) is not a solution to the problem (c) (see Fig. 14.1).

In such a situation, it is reasonable to consider multi-stage optimization of matchings relative to different criteria, for example, to describe all matchings with the maximum cardinality and later to describe among these matchings all matchings with the maximum total weight of edges. We can also describe the whole set of matchings with the maximum total weight of nodes and after that can describe among these matchings all matchings with the maximum total weight of edges.

M. Mankowski and M. Moshkov, *Dynamic Programming Multi-Objective Combinatorial Optimization*, Studies in Systems, Decision and Control 331,
https://doi.org/10.1007/978-3-030-63920-4_14

Fig. 14.1 Matching with the maximum weight of edges (**a**) and matching with the maximum cardinality (**b**)

Another possibility is to consider bi-criteria optimization of matchings relative to the cardinality of matching and the sum of weights of edges in matching. We can also consider the sum of weights of edges in matching and the sum of weights of nodes in matching. All these criteria should be maximized.

The result of bi-criteria optimization is the set of Pareto optimal points (POPs). If we consider bi-criteria optimization relative to the cardinality of matching and the sum of weights of nodes in matching, we will have only one POP that corresponds to matchings with the maximum sum of weights of nodes and with the maximum cardinality (it follows from results mentioned in [2, 4, 8]). We call such matchings, which are optimal relative to two or more criteria simultaneously, totally optimal matchings relative to these criteria.

In the case of bi-criteria optimization relative to the cardinality of matching and the sum of weights of edges in matching, we can have more than one POP. For example, for the problem presented in Fig. 14.1, there are exactly two POPs (1, 3) and (2, 2) corresponding to matchings (a) and (b), respectively. The first coordinate is the cardinality of matching, and the second one—the sum of weights of edges in matching.

The problem (c) can be formulated as the problem (e) when the weight of each edge is equal to 1. The problem (n) can be formulated as the problem (e) when the weight of each edge is equal to the sum of weights of its ends. So we can consider multi-stage optimization of matchings relative to several weight functions each of which assigns a positive integer (weight) to each edge of the graph G. The same situation is with bi-criteria optimization of matchings.

In this part, we create dynamic programming algorithms (i) for counting the matchings in trees before and after optimization, (ii) for multi-stage optimization of matchings in trees relative to a sequence of weight functions, and (iii) for the bi-criteria optimization of matchings in trees relative to two weight functions. The algorithms (i) and (ii) require a linear number of arithmetical operations depending on the number of nodes in the tree. The algorithm (iii) requires a polynomial number of arithmetical operations depending on the number of nodes in the tree and the maximum weight of an edge in the tree for the first weight function.

We implemented the considered algorithms and made computer experiments to evaluate the number of matchings before and after optimization, to compare the maximum cardinality of matchings before and after optimization related to a weight function, to evaluate the number of POPs, and to study the existence of totally optimal matchings relative to the cardinality of matching and to the total weight of edges in matching.

To the best of our knowledge, there are no previously created polynomial exact algorithms for multi-stage and bi-criteria optimization of matchings for trees and more general classes of graphs, which outcome all possible optimal solutions in each stage of optimization or create the set of Pareto optimal points.

This part is an essentially extended version of conference papers [7] and [6] describing multi-stage and bi-criteria optimization of matchings in trees. These two papers were inspired by problems of matching optimization related to kidney paired donation [2, 3, 8, 10]. This is a novel alternative for living, incompatible (donor, recipient) pairs to get an organ by matching with another incompatible pair.

Let G be an undirected graph in which edges and nodes have positive weights. Nodes of this graph are pairs (donor, recipient) and two nodes $A = (a_1, a_2)$ and $B = (b_1, b_2)$ are connected by an edge if the donor a_1 can donate a kidney to the recipient b_2, and the donor b_1 can donate a kidney to the recipient a_2. The weight of a node A is the importance of the transplantation for the recipient from the pair A. The weight of an edge connecting nodes A and B is the compatibility of the exchange of kidneys between the pairs A and B.

Each solution to the problem (c) for the graph G allows us to help the maximum number of patients. As we already mentioned, each solution to the problem (n)—maximization of total importance of transplantations—is also a solution to the problem (c). However, it is possible that a solution to the problem (e)—maximization of the total compatibility of exchanges—is not a solution to the problem (c). This part shows possible ways to deal with this issue in the case when the graph G is a tree. To apply to real kidney paired donation problems, the considered approaches should be extended to arbitrary undirected graphs.

14.2 Representation of Matchings in Trees

Let G be a tree. A *matching* in G is a set of edges without common nodes. We choose a node r in the tree G as the root. It will be useful for us to consider G as a directed graph with the orientation of edges from the root r. Now each node v in G defines a subtree $G(v)$ of G in which v is the root.

We now describe a graph $D(G)$ (labeled forest), which will be used to describe the set of matchings in G and to optimize these matchings. It contains the main nodes from G and auxiliary nodes corresponding to the main ones.

Let v be a terminal node of G—see Fig. 14.2a. Then in the graph $D(G)$, there are two nodes v (main) and $v(\emptyset)$ (auxiliary) corresponding to v, which are connected by an edge leaving v and entering $v(\emptyset)$—see Fig. 14.2b.

Let v be a nonterminal node of G, which has k leaving edges $e_1, \ldots, e_k$ entering nodes $v_1, \ldots, v_k$, respectively—see Fig. 14.3a. Then in $D(G)$ there are the main node v, $k + 1$ auxiliary nodes $v(e_1), \ldots, v(e_k)$, $v(\emptyset)$, and $k + 1$ edges leaving v and entering these auxiliary nodes—see Fig. 14.3b.

An *edge-preserving subgraph* Δ of the graph $D(G)$ is a graph obtained from $D(G)$ by the removal of some edges such that each main node in Δ has at least one

Fig. 14.2 Nodes and edge in $D(G)$ (**b**) corresponding to terminal node v of G (**a**)

Fig. 14.3 Nodes and edges in $D(G)$ (**b**) corresponding to nonterminal node v of G (**a**)

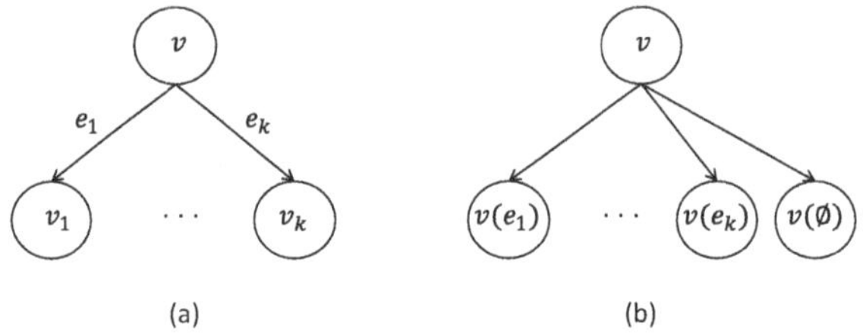

leaving edge. Let v be the main node in $D(G)$ with children $v(e_1), \ldots, v(e_k), v(\emptyset)$. We denote by $E_\Delta(v)$ the set of all $\sigma \in \{e_1, \ldots, e_k, \emptyset\}$ such that there is an edge in Δ from v to $v(\sigma)$. We assume that $D(G)$ is an edge-preserving subgraph of $D(G)$. Edge-preserving subgraphs of the graph $D(G)$ can be obtained as results of optimization of matchings in G relative to weight functions.

Let Δ be an edge-preserving subgraph of the graph $D(G)$. We correspond a set $M_\Delta(u)$ of matchings in G to each node u of Δ. Let C and D be nonempty sets of matchings in G. We denote $C \otimes D = \{c \cup d : c \in C, d \in D\}$.

Let v be a terminal node of G. Then

$$M_\Delta(v) = M_\Delta(v(\emptyset)) = \{\lambda\} , \tag{14.1}$$

where λ is the empty matching. Let v be a nonterminal node of G, which has k leaving edges $e_1, \ldots, e_k$ entering nodes $v_1, \ldots, v_k$, respectively. Then, for $i = 1, \ldots, k$,

$$M_\Delta(v(\emptyset)) = \bigotimes_{j \in \{1,\ldots,k\}} M_\Delta(v_j) , \tag{14.2}$$

$$M_\Delta(v(e_i)) = \left(\bigotimes_{j \in \{1,\ldots,k\} \setminus \{i\}} M_\Delta(v_j) \right) \otimes M_\Delta(v_i(\emptyset)) \otimes \{\{e_i\}\} , \tag{14.3}$$

$$M_\Delta(v) = \bigcup_{\sigma \in E_\Delta(v)} M_\Delta(v(\sigma)) . \tag{14.4}$$

Let G contain n nodes and, therefore, $n - 1$ edges. Then the graph $D(G)$ contains $3n - 1$ nodes and $2n - 1$ edges. It is clear that the graph $D(G)$ can be constructed in linear time depending on n.

One can prove the following statement related to the case when $\Delta = D(G)$.

Proposition 14.1 *Let G be a tree with root. Then, for any node v of G, $M_{D(G)}(v)$ is the set of all matchings in $G(v)$ and $M_{D(G)}(v(\emptyset))$ is the set of all matchings in $G(v)$, which have no edges with the end v. For any nonterminal node v of G and for any edge e leaving v, $M_{D(G)}(v(e))$ is the set of all matchings in $G(v)$ containing the edge e.*

Let G be a tree and A, B be nonempty sets of matchings in G. We will say that A and B are *disjoint* if $A \cap B = \emptyset$. We will say that A and B are *strongly disjoint* if $a \cap b = \emptyset$ for any $a \in A$ and $b \in B$.

Lemma 14.1 *Let G be a tree with root and Δ be an edge-preserving subgraph of the graph $D(G)$. Let v be a nonterminal node of G, which has k leaving edges $e_1, \ldots, e_k$ entering nodes $v_1, \ldots, v_k$, respectively and, for $j = 1, \ldots, k$, $A_j \in \{M_\Delta(v_j), M_\Delta(v_j(\emptyset))\}$. Then the sets $A_1, \ldots, A_k$ are pairwise strongly disjoint and the sets $M_\Delta(v(e_1)), \ldots, M_\Delta(v(e_k)), M_\Delta(v(\emptyset))$ are pairwise disjoint.*

Proof Let $i, j \in \{1, \ldots, k\}$, $i \neq j$, $a_i \in A_i$ and $a_j \in A_j$. Then a_i is a matching in the subtree $G(v_i)$ and a_j is a matching in the subtree $G(v_j)$ of the tree G. It is clear that these subtrees have no common edges. Therefore, $a_i \cap a_j = \emptyset$ and sets A_i and A_j are strongly disjoint. Hence, the sets $A_1, \ldots, A_k$ are pairwise strongly disjoint.

Let $i, j \in \{1, \ldots, k\}$, $i \neq j$, $a_i \in M_\Delta(v(e_i))$, $a_j \in M_\Delta(v(e_j))$, and $a_0 \in M_\Delta(v(\emptyset))$. It is clear that a_i contains the edge e_i but a_j and a_0 do not contain e_i. Therefore, $a_i \neq a_j$ and $a_i \neq a_0$. From here, it follows that the sets $M_\Delta(v(e_1)), \ldots, M_\Delta(v(e_k)), M_\Delta(v(\emptyset))$ are pairwise disjoint. □

References

1. Edmonds, J.: Paths, trees, and flowers. Canad. J. Math. **17**, 449–467 (1965)
2. Gentry, S., Mankowski, M., Michael, T.: Maximum matchings in graphs for allocating kidney paired donation. Oper. Res. Health Care **25** (2020)
3. Gentry, S.E.: Optimization over graphs for kidney paired donation. In: Lim, G.J., Lee, E.K. (eds.) Optimization in Medicine and Biology, pp. 177–195. Auerbach Publications, Taylor & Francis Group (2008)
4. Halappanavar, M.: Algorithms for vertex-weighted matching in graphs. Ph.D. thesis, Old Dominion University, Norfolk, VA, USA (2009)
5. Lovász, L., Plummer, M.D.: Matching Theory. Akadémiai Kiadó—North Holland (1986)
6. Mankowski, M., Moshkov, M.: Bi-criteria optimization of matchings in trees with application to kidney exchange. In: 26th International Workshop on Concurrency, Specification and Programming, CS&P 2017, Warsaw, Poland, September 25–27, 2017 (2017). http://csp2017.mimuw.edu.pl/data/uploads/papers/CSP2017_paper_1.pdf
7. Mankowski, M., Moshkov, M.: Multi-stage optimization of matchings in trees with application to kidney exchange. In: Polkowski, L., Yao, Y., Artiemjew, P., Ciucci, D., Liu, D., Slezak, D., Zielosko, B. (eds.) Proceedings of the International Joint Conference on Rough Sets (IJCRS 2017), Part I, Olsztyn, Poland, July 3–7, 2017. Lecture Notes in Computer Science, vol. 10313, pp. 123–130. Springer (2017)
8. Roth, A., Sönmez, T., Unver, U.: Pairwise kidney exchange. J. Econom. Theory **125**(2), 151–188 (2005)

9. Savage, C.D.: Maximum matchings and trees. Inf. Process. Lett. **10**(4/5), 202–205 (1980)
10. Segev, D.L., Gentry, S.E., Warren, D.S., Reeb, B., Montgomery, R.A.: Kidney paired donation and optimizing the use of live donor organs. JAMA J. Am. Med. Assoc. **293**(15), 1883–1890 (2005)

Chapter 15
Counting Matchings and Multi-Stage Optimization of Matchings

In this chapter, we consider the algorithm $\mathscr{A}_5$ for counting matchings and the algorithm $\mathscr{A}_6$ for optimization of matchings relative to a weight function. The algorithm of optimization $\mathscr{A}_6$ can be used for multi-stage optimization of matchings relative to a sequence of weight functions, and the algorithm $\mathscr{A}_5$ can be used for counting the optimal matchings after each step of optimization. This chapter contains some revised results from the conference paper [1].

15.1 Counting Matchings

Let G be a tree with the root r and Δ be an edge-preserving subgraph of the graph $D(G)$. The graph Δ contains main nodes from G and auxiliary nodes corresponding to the main ones. We corresponded a set $M_\Delta(u)$ of matchings in G to each node u of Δ. We now describe an algorithm, which computes, for each node u of the graph Δ, the cardinality $C(u)$ of the set $M_\Delta(u)$, and returns the number $C(r) = |M_\Delta(r)|$.

Algorithm $\mathscr{A}_5$ (counting matchings)

Input: A tree G with the root r and an edge-preserving subgraph Δ of the graph $D(G)$.
Output: The number $C(r)$.

1. If all main nodes of the graph Δ are processed, then return the number $C(r)$ and finish the work of the algorithm. Otherwise, choose a main node v of the graph Δ, which is not processed yet and which is either a terminal node of G or a nonterminal node of G such that all children of v in G are processed.
2. Let v be a terminal node of G. Then set

M. Mankowski and M. Moshkov, *Dynamic Programming Multi-Objective Combinatorial Optimization*, Studies in Systems, Decision and Control 331,
https://doi.org/10.1007/978-3-030-63920-4_15

$$C(v) = C(v(\emptyset)) = 1\ , \tag{15.1}$$

mark the node v as processed, and proceed to step 1.

3. Let v be a nonterminal node of G, which has k leaving edges $e_1, \ldots, e_k$ entering nodes $v_1, \ldots, v_k$, respectively. Then, for $i = 1, \ldots, k$, set

$$C(v(\emptyset)) = \prod_{j \in \{1,\ldots,k\}} C(v_j)\ , \tag{15.2}$$

$$\begin{aligned} C(v(e_i)) &= \left(\prod_{j \in \{1,\ldots,k\} \setminus \{i\}} C(v_j) \right) \cdot C(v_i(\emptyset)) \\ &= \frac{C(v(\emptyset)) \cdot C(v_i(\emptyset))}{C(v_i)}\ , \end{aligned} \tag{15.3}$$

$$C(v) = \sum_{\sigma \in E_\Delta(v)} C(v(\sigma))\ , \tag{15.4}$$

mark the node v as processed, and proceed to step 1.

First, we prove the following auxiliary statement.

Lemma 15.1 *Let G be a tree and $A_1, \ldots, A_k$ be nonempty sets of matchings in G. If $A_k, \ldots, A_k$ are pairwise disjoint, then $|A_1 \cup \cdots \cup A_k| = |A_1| + \cdots + |A_k|$. If $A_k, \ldots, A_k$ are pairwise strongly disjoint, then $|A_1 \otimes \cdots \otimes A_k| = |A_1| \cdot \ldots \cdot |A_k|$.*

Proof The first statement is trivial. Let us prove the second one. Let $a_1, b_1 \in A_1, \ldots, a_k, b_k \in A_k$ and, for some $i \in \{1, \ldots, k\}$, $a_i \neq b_i$. It is clear that there is an edge $e \in a_i \cup b_i$ such that $e \notin a_i \cap b_i$. Since $A_1, \ldots, A_k$ are pairwise strongly disjoint, $e \notin \bigcup_{j \in \{1,\ldots,k\} \setminus \{i\}} (a_j \cup b_j)$. Therefore, $a_1 \cup \cdots \cup a_k \neq b_1 \cup \cdots \cup b_k$. Using this fact, it is easy to show that $|A_1 \otimes \cdots \otimes A_k| = |A_1| \cdot \ldots \cdot |A_k|$. □

We now prove the correctness of the algorithm $\mathscr{A}_5$ and analyze its time complexity.

Theorem 15.1 *Let G be a tree with the root r containing n nodes and Δ be an edge-preserving subgraph of the graph $D(G)$. Then the algorithm $\mathscr{A}_5$ returns the number $C(r) = |M_\Delta(r)|$ and makes $O(n)$ operations of addition, multiplication, and division.*

Proof We prove by induction on the nodes of Δ that $C(u) = |M_\Delta(u)|$ for each node u of Δ. Let v be a terminal node of G. Then by (14.1), $M_\Delta(v) = M_\Delta(v(\emptyset)) = \{\lambda\}$ and $|M_\Delta(v)| = |M_\Delta(v(\emptyset))| = 1$. Using (15.1) we obtain that the considered statement holds for v and $v(\emptyset)$.

Let now v be a nonterminal node of G, which has k leaving edges $e_1, \ldots, e_k$ entering nodes $v_1, \ldots, v_k$, respectively, and, for nodes $v_1, \ldots, v_k, v_1(\emptyset), \ldots, v_k(\emptyset)$, the considered statement hold. Let us show that, for $i = 1, \ldots, k$,

$$|M_\Delta(v(\emptyset))| = \prod_{j\in\{1,\dots,k\}} \left|M_\Delta(v_j)\right| , \tag{15.5}$$

$$\begin{aligned}|M_\Delta(v(e_i))| &= \left(\prod_{j\in\{1,\dots,k\}\setminus\{i\}} \left|M_\Delta(v_j)\right|\right) \cdot |M_\Delta(v_i(\emptyset))| \\ &= \frac{|M_\Delta(v(\emptyset))| \cdot |M_\Delta(v_i(\emptyset))|}{|M_\Delta(v_i)|} ,\end{aligned} \tag{15.6}$$

$$|M_\Delta(v)| = \sum_{\sigma\in E_\Delta(v)} |M_\Delta(v(\sigma))| . \tag{15.7}$$

From (14.2) and Lemmas 14.1 and 15.1 it follows that the equality (15.5) holds. Using (14.3) we obtain

$$|M_\Delta(v(e_i))| = \left|\left(\bigotimes_{j\in\{1,\dots,k\}\setminus\{i\}} M_\Delta(v_j)\right) \otimes M_\Delta(v_i(\emptyset))\right| .$$

By Lemmas 14.1 and 15.1,

$$|M_\Delta(v(e_i))| = \left(\prod_{j\in\{1,\dots,k\}\setminus\{i\}} \left|M_\Delta(v_j)\right|\right) \cdot |M_\Delta(v_i(\emptyset))| .$$

From here and from (15.5) it follows that the equality (15.6) holds. The equality (15.7) follows from Lemmas 14.1 and 15.1.

By the inductive hypothesis, $C(v_i) = |M_\Delta(v_i))|$ and $C(v_i(\emptyset)) = |M_\Delta(v_i(\emptyset)))|$ for $i = 1, \dots, k$. Using (15.2)–(15.4) and (15.5)–(15.7) we obtain that $C(v(\emptyset)) = |M_\Delta(v(\emptyset))|, C(v(e_i)) = |M_\Delta(v(e_i))|$ for $i = 1, \dots, k$, and $C(v) = |M_\Delta(v)|$. Hence, the considered statement holds for v, $v(\emptyset)$, $v(e_1), \dots, v(e_k)$. We proved that $C(u) = |M_\Delta(u)|$ for any node u of the graph $D(G)$. From here, it follows that $C(r) = |M_\Delta(r)|$, and the algorithm $\mathscr{A}_5$ returns the cardinality of the set $M_\Delta(r)$.

Since G is a tree, the number of edges in G is equal to $n - 1$. For a terminal node v of the tree G, we do not need arithmetical operations to find values of $C(v)$ and $C(v(\emptyset))$—see (15.1). Let v be a nonterminal node of G, which has k leaving edges $e_1, \dots, e_k$ entering nodes $v_1, \dots, v_k$, respectively. We need $k - 1$ multiplications to compute the value $C(v(\emptyset))$, $2k$ multiplications and divisions to compute the values $C(v(e_1)), \dots, C(v(e_k))$, and at most k additions to compute the value $C(v)$—see (15.2)–(15.4). As a result, for the node v, the considered algorithm makes at most $4k$ arithmetical operations. To process the tree G, the algorithm $\mathscr{A}_5$ makes at most $4n$ operations of addition, multiplication, and division. □

15.2 Multi-Stage Optimization of Matchings

Let G be a tree with the root r, Δ be an edge-preserving subgraph of the graph $D(G)$, and w be a weight function, which assigns a positive integer (weight) $w(e)$ to each edge e of G. We now describe the procedure of optimization of matchings represented by Δ relative to the weight function w. During the work of this procedure, we attach a number $g(u)$ to each node u of Δ and, maybe, remove some edges from Δ. As a result, we obtain an edge-preserving subgraph Δ^w of the graph Δ. It is clear that Δ^w is also an edge-preserving subgraph of the graph $D(G)$. We will show that, for each node u of the graph Δ, the number $g(u)$ is the maximum total weight of edges in a matching from $M_\Delta(u)$ relative to the weight function w, and the set $M_{\Delta^w}(u)$ is the set of all matchings from $M_\Delta(u)$ that have the total weight of edges $g(u)$ relative to the weight function w.

Algorithm $\mathscr{A}_6$ (procedure of matching optimization)

Input: A tree G with root, an edge-preserving subgraph Δ of the graph $D(G)$, and a weight function w.
Output: The edge-preserving subgraph Δ^w of the graph Δ.

1. If all main nodes of the graph Δ are processed, then return the obtained graph as Δ^w and finish the work of the algorithm. Otherwise, choose a main node v of the graph Δ, which is not processed yet and which is either a terminal node of G or a nonterminal node of G such that all children of v in G are processed.
2. Let v be a terminal node of G. Then set
$$g(v) = g(v(\emptyset)) = 0\ , \tag{15.8}$$
mark node v as processed and proceed to step 1.
3. Let v be a nonterminal node of G, which has k leaving edges $e_1, \ldots, e_k$ entering nodes $v_1, \ldots, v_k$, respectively. Then, for $i = 1, \ldots, k$, set
$$g(v(\emptyset)) = \sum_{j\in\{1,\ldots,k\}} g(v_j)\ , \tag{15.9}$$
$$\begin{aligned} g(v(e_i)) &= \left(\sum_{j\in\{1,\ldots,k\}\setminus\{i\}} g(v_j)\right) + g(v_i(\emptyset)) + w(e_i) \\ &= g(v(\emptyset)) - g(v_i) + g(v_i(\emptyset)) + w(e_i)\ , \end{aligned} \tag{15.10}$$
$$g(v) = \max\{g(v(\sigma)) : \sigma \in E_\Delta(v)\}\ . \tag{15.11}$$
For each $\sigma \in E_\Delta(v)$ such that $g(v(\sigma)) < g(v)$, remove the edge connecting nodes v and $v(\sigma)$ from the graph Δ. Mark the node v as processed and proceed to step 1.

Note that in the obtained edge-preserving subgraph Δ^w of the graph $D(G)$, for each nonterminal node v of the tree G,

$$E_{\Delta^w}(v) = \{\sigma : \sigma \in E_\Delta(v), g(v(\sigma)) = g(v)\} . \tag{15.12}$$

Let a be a matching in the graph G and A be a nonempty set of matchings in G. We denote $w(a) = \sum_{e \in a} w(e)$ (in particular, $w(\lambda) = 0$, where λ is the empty matching), $w^{opt}(A) = \max\{w(a) : a \in A\}$ and $Opt_w(A) = \{a \in A : w(a) = w^{opt}(A)\}$. It is not difficult to prove the following statement.

Lemma 15.2 *Let G be a tree with root, $A_1, \ldots, A_k$ be nonempty sets of matchings in G, and $A = A_1 \cup \cdots \cup A_k$. Then $w^{opt}(A) = \max\{w^{opt}(A_1), \ldots, w^{opt}(A_k)\}$ and $Opt_w(A) = \bigcup_{\substack{i \in \{1,\ldots,k\}, \\ w^{opt}(A_i) = w^{opt}(A)}} Opt_w(A_i)$.*

Lemma 15.3 *Let G be a tree with root, $A_1, \ldots, A_k$ be pairwise strongly disjoint nonempty sets of matchings in G, and $A = A_1 \otimes \cdots \otimes A_k$. Then $w^{opt}(A) = w^{opt}(A_1) + \cdots + w^{opt}(A_k)$ and $Opt_w(A) = Opt_w(A_1) \otimes \cdots \otimes Opt_w(A_k)$.*

Proof Let $a_1 \in A_1, \ldots, a_k \in A_k$. It is clear that $w(a_1 \cup \cdots \cup a_k) = w(a_1) + \cdots + w(a_k)$. Using this fact it is easy to show that $w^{opt}(A) = w^{opt}(A_1) + \cdots + w^{opt}(A_k)$ and $Opt_w(A) = Opt_w(A_1) \otimes \cdots \otimes Opt_w(A_k)$. □

Theorem 15.2 *Let G be a tree with root containing n nodes, Δ be an edge-preserving subgraph of the graph $D(G)$, and w be a weight function. Then the algorithm $\mathscr{A}_6$ returns the edge-preserving subgraph Δ^w of the graph Δ such that $g(u) = w^{opt}(M_\Delta(u))$ and $M_{\Delta^w}(u) = Opt_w(M_\Delta(u))$ for any node u of the graph Δ, and makes $O(n)$ additions, subtractions, and comparisons.*

Proof We prove by induction on the nodes of Δ that $g(u) = w^{opt}(M_\Delta(u))$ and $M_{\Delta^w}(u) = Opt_w(M_\Delta(u))$ for each node u of Δ. Let v be a terminal node of G. Using (14.1) we obtain that $M_\Delta(v) = M_\Delta(v(\emptyset)) = M_{\Delta^w}(v) = M_{\Delta^w}(v(\emptyset)) = \{\lambda\}$, $w^{opt}(M_\Delta(v)) = w^{opt}(M_\Delta(v(\emptyset))) = 0$, and $Opt_w(M_\Delta(v)) = Opt_w(M_\Delta(v(\emptyset))) = \{\lambda\}$. From these equalities and from (15.8) it follows that the considered statement holds for v and $v(\emptyset)$.

Let now v be a nonterminal node of G, which has k leaving edges $e_1, \ldots, e_k$ entering nodes $v_1, \ldots, v_k$, respectively, and, for nodes $v_1, \ldots, v_k, v_1(\emptyset), \ldots, v_k(\emptyset)$, the considered statement hold. Let us show that, for $i = 1, \ldots, k$,

$$w^{opt}(M_\Delta(v(\emptyset))) = \sum_{j \in \{1,\ldots,k\}} w^{opt}(M_\Delta(v_j)) , \tag{15.13}$$

$$\begin{aligned} w^{opt}(M_\Delta(v(e_i))) &= \left(\sum_{j \in \{1,\ldots,k\} \setminus \{i\}} w^{opt}(M_\Delta(v_j)) \right) \\ &\quad + w^{opt}(M_\Delta(v_i(\emptyset))) + w(e_i) \\ &= w^{opt}(M_\Delta(v(\emptyset))) - w^{opt}(M_\Delta(v_i)) \\ &\quad + w^{opt}(M_\Delta(v_i(\emptyset))) + w(e_i) , \end{aligned} \tag{15.14}$$

$$w^{opt}(M_\Delta(v)) = \max\{w^{opt}(M_\Delta(v(\sigma))) : \sigma \in E_\Delta(v)\} . \tag{15.15}$$

The equality (15.13) follows from (14.2) and Lemmas 14.1 and 15.3. Using Lemmas 14.1 and 15.3 we obtain that

$$w^{opt}\left(\left(\bigotimes_{j\in\{1,\ldots,k\}\setminus\{i\}} M_{\Delta}(v_j)\right)\otimes M_{\Delta}(v_i(\emptyset))\right)$$
$$=\left(\sum_{j\in\{1,\ldots,k\}\setminus\{i\}} w^{opt}(M_{\Delta}(v_j))\right)+w^{opt}(M_{\Delta}(v_i(\emptyset)))\ .$$

Using this equality and equalities (14.3) and (15.13) one can show that the equality (15.14) holds. The equality (15.15) follows from the equality (14.4) and Lemma 15.2. By the inductive hypothesis, $g(v_1)=w^{opt}(M_{\Delta}(v_1)),\ldots,g(v_k)=w^{opt}(M_{\Delta}(v_k))$, $g(v_1(\emptyset))=w^{opt}(M_{\Delta}(v_1(\emptyset))),\ldots,g(v_k(\emptyset))=w^{opt}(M_{\Delta}(v_k(\emptyset))$. Using the equalities (15.9)–(15.11) and (15.13)–(15.15) we obtain $g(v(\emptyset))=w^{opt}(M_{\Delta}(v(\emptyset)))$, $g(v(e_i))=w^{opt}(M_{\Delta}(v(e_i)))$ for $i=1,\ldots,k$, and $g(v)=w^{opt}(M_{\Delta}(v))$.

Let us show that, for $i=1,\ldots,k$,

$$Opt_w(M_{\Delta}(v(\emptyset)))=\bigotimes_{j\in\{1,\ldots,k\}} Opt_w(M_{\Delta}(v_j))\ , \tag{15.16}$$

$$Opt_w(M_{\Delta}(v(e_i)))=\left(\bigotimes_{j\in\{1,\ldots,k\}\setminus\{i\}} Opt_w(M_{\Delta}(v_j))\right)$$
$$\otimes Opt_w(M_{\Delta}(v_i(\emptyset)))\otimes\{\{e_i\}\}\ , \tag{15.17}$$

$$Opt_w(M_{\Delta}(v))=\bigcup_{\substack{\sigma\in E_{\Delta}(v),\\ w^{opt}(M_{\Delta}(v(\sigma)))=w^{opt}(M_{\Delta}(v))}} Opt_w(M_{\Delta}(v(\sigma)))\ . \tag{15.18}$$

The equality (15.16) follows from the equality (14.2) and Lemmas 14.1 and 15.3. Using Lemma 14.1 one can show that the sets

$$M_{\Delta}(v_1),\ldots,M_{\Delta}(v_{i-1}),M_{\Delta}(v_i(\emptyset)),M_{\Delta}(v_{i+1}),\ldots,M_{\Delta}(v_k),\{\{e_i\}\}$$

are pairwise strongly disjoint. It is clear that $Opt_w(\{\{e_i\}\})=\{\{e_i\}\}$. From here, from (14.3), and from Lemma 15.3 it follows that the equality (15.17) holds. The equality (15.18) follows from (14.4) and Lemma 15.2.

By the inductive hypothesis, $M_{\Delta^w}(u)=Opt_w(M_{\Delta}(u))$ for any

$$u\in\{v_1,\ldots,v_k,v_1(\emptyset),\ldots,v_k(\emptyset)\}\ .$$

Using (15.16)–(15.18) we obtain that, for $i=1,\ldots,k$,

$$Opt_w(M_\Delta(v(\emptyset))) = \bigotimes_{j\in\{1,\ldots,k\}} M_{\Delta^w}(v_j) \; , \tag{15.19}$$

$$Opt_w(M_\Delta(v(e_i))) = \left(\bigotimes_{j\in\{1,\ldots,k\}\setminus\{i\}} M_{\Delta^w}(v_j)\right) \otimes M_{\Delta^w}(v_i(\emptyset)) \otimes\{\{e_i\}\} \; , \tag{15.20}$$

$$Opt_w(M_\Delta(v)) = \bigcup_{\substack{\sigma\in E_\Delta(v),\\ w^{opt}(M_\Delta(v(\sigma)))=w^{opt}(M_\Delta(v))}} M_{\Delta^w}(v(\sigma)) \; . \tag{15.21}$$

We know (see (14.2)–(14.4)) that, for $i = 1, \ldots, k$,

$$M_{\Delta^w}(v(\emptyset)) = \bigotimes_{j\in\{1,\ldots,k\}} M_{\Delta^w}(v_j) \; , \tag{15.22}$$

$$M_{\Delta^w}(v(e_i)) = \left(\bigotimes_{j\in\{1,\ldots,k\}\setminus\{i\}} M_{\Delta^w}(v_j)\right) \otimes M_{\Delta^w}(v_i(\emptyset)) \otimes \{\{e_i\}\} \; , \tag{15.23}$$

$$M_{\Delta^w}(v) = \bigcup_{\sigma\in E_{\Delta^w}(v)} M_{\Delta^w}(v(\sigma)) \; . \tag{15.24}$$

From (15.19) and (15.22) it follows that $Opt_w(M_\Delta(v(\emptyset))) = M_{\Delta^w}(v(\emptyset))$. From (15.20) and (15.23) it follows that $Opt_w(M_\Delta(v(e_i))) = M_{\Delta^w}(v(e_i))$ for $i = 1, \ldots, k$. We proved that $w^{opt}(M_\Delta(v(\sigma))) = g(v(\sigma))$ for any $\sigma \in E_\Delta(v)$ and $w^{opt}(M_\Delta(v)) = g(v)$. From here, (15.12) and equalities (15.21), (15.24) it follows $Opt_w(M_\Delta(v)) = M_{\Delta^w}(v)$. Therefore, the considered statement holds also for nodes $v(\emptyset), v(e_1), \ldots, v(e_k), v$. Hence, $g(u) = w^{opt}(M_\Delta(u))$ and $M_{\Delta^w}(u) = Opt_w(M_\Delta(u))$ for any node u of the graph Δ.

The tree G contains $n - 1$ edges. For a terminal node v of the tree G, we do not need arithmetical operations to find values of $g(v)$ and $g(v(\emptyset))$ (see (15.8)). Let v be a nonterminal node of G, which has k leaving edges $e_1, \ldots, e_k$ entering nodes $v_1, \ldots, v_k$, respectively. We need $k - 1$ additions to compute the value $g(v(\emptyset))$, $3k$ additions and subtractions to compute the values $g(v(e_1)), \ldots, g(v(e_k))$, at most k comparisons to compute the value $g(v)$, and at most $k + 1$ comparisons to determine edges starting in v that should be removed—see (15.9)–(15.11) and the last part of step 3. As a result, for the node v, the considered algorithm makes at most $6k$ arithmetical operations. To process the tree G, the algorithm $\mathscr{A}_6$ makes at most $6n$ arithmetical operations (additions, subtractions, and comparisons). □

We can use the considered procedure (algorithm $\mathscr{A}_6$) for multi-stage optimization of matchings. Let us have a tree G and weight functions $w_1, w_2, \ldots$, which assign positive integers (weights) to the edges of G. We choose a node r of G as the root and construct the graph $\Delta = D(G)$. We know that the set $M_\Delta(r)$ corresponding to the node r of Δ is equal to the set of all matchings in G.

We apply to the graph Δ the procedure of optimization relative to the weight function w_1. As a result, we obtain an edge-preserving subgraph Δ^{w_1} of the graph $D(G)$. The set $M_{\Delta^{w_1}}(r)$ corresponding to the node r of Δ^{w_1} is equal to the set of all matchings from $M_{\Delta}(r)$, which have the maximum total weight of edges relative to the weight function w_1.

We apply to the graph Δ^{w_1} the procedure of optimization relative to the weight function w_2. As a result, we obtain an edge-preserving subgraph Δ^{w_1,w_2} of the graph $D(G)$. The set of matchings $M_{\Delta^{w_1,w_2}}(r)$ corresponding to the node r of Δ^{w_1,w_2} is equal to the set of all matchings from $M_{\Delta^{w_1}}(v)$, which have the maximum total weight of edges relative to the weight function w_2, etc.

In particular, we can maximize the cardinality of matchings and after that among all matchings with the maximum cardinality, we can choose all matchings with the maximum total compatibility [1].

Reference

1. Mankowski, M., Moshkov, M.: Multi-stage optimization of matchings in trees with application to kidney exchange. In: Polkowski, L., Yao, Y., Artiemjew, P., Ciucci, D., Liu, D., Slezak, D., Zielosko, B. (eds.) Proceedings of the International Joint Conference on Rough Sets (IJCRS 2017), Part I, Olsztyn, Poland, July 3–7, 2017. Lecture Notes in Computer Science, vol. 10313, pp. 123–130. Springer (2017)

Chapter 16
Bi-Criteria Optimization of Matchings

In this chapter, we describe an additional tool for the study of Pareto optimal points (POPs) in comparison to ones considered in Sect. 4.1: the algorithm $\mathscr{A}_7$ for the fusion of sets of POPs [1]. We also propose the algorithm $\mathscr{A}_8$, which constructs the set of POPs for bi-criteria optimization of matchings in trees relative to two weight functions. In the end of the chapter, we discuss the notion of a totally optimal matching (optimal relative to two weight functions simultaneously) and show how we can recognize the existence of totally optimal matchings using multi-stage and bi-criteria optimization algorithms. This chapter contains some revised results from the conference paper [4].

16.1 Fusion of Sets of Pareto Optimal Points

Let A, B be nonempty finite subsets of the set $\mathbb{R}^2$. We denote by $A \oplus B$ the set $\{(a + c, b + d) : (a, b) \in A, (c, d) \in B\}$. Note that $A \oplus B = J_{f,g}(A, B)$, where $f = g = x_1 + x_2$ (see Sect. 4.1).

Let $P_1, \ldots, P_t$ be nonempty finite subsets of $\mathbb{R}^2$, $Q_1 = P_1$, and, for $i = 2, \ldots, t$, $Q_i = Q_{i-1} \oplus P_i$. We assume that, for $i = 1, \ldots, t$, the sets $Par(P_1), \ldots, Par(P_t)$ are already constructed. We now describe an algorithm that constructs the sets $Par(Q_1), \ldots, Par(Q_t)$ and returns $Par(Q_t) = Par(P_1 \oplus \cdots \oplus P_t)$. This algorithm is a special case of a more general algorithm considered in [1].

Algorithm $\mathscr{A}_7$ (fusion of sets of POPs).

Input: Sets $Par(P_1), \ldots, Par(P_t)$ for some nonempty finite subsets $P_1, \ldots, P_t$ of $\mathbb{R}^2$.

Output: The set $Par(Q_t)$, where $Q_1 = P_1$, and, for $i = 2, \ldots, t$, $Q_i = Q_{i-1} \oplus P_i$, i.e., $Q_t = P_1 \oplus \cdots \oplus P_t$.

M. Mankowski and M. Moshkov, *Dynamic Programming Multi-Objective Combinatorial Optimization*, Studies in Systems, Decision and Control 331,
https://doi.org/10.1007/978-3-030-63920-4_16

1. Set $B_1 = Par(P_1)$ and set $i = 2$.
2. Construct the multiset

$$A_i = B_{i-1} \oplus Par(P_i) = \{(a+c, b+d)) : (a,b) \in B_{i-1}, (c,d) \in Par(P_i)\}$$

   (we will not remove equal pairs from the constructed set).
3. Using algorithm $\mathscr{A}_3$, construct the set $B_i = Par(A_i)$.
4. If $i = t$, then return B_i and finish the work of the algorithm. Otherwise, set $i = i + 1$ and proceed to step 2.

Proposition 16.1 [1] *Let $P_1, \ldots, P_t$ be nonempty finite subsets of $\mathbb{R}^2$, $Q_1 = P_1$, and, for $i = 2, \ldots, t$, $Q_i = Q_{i-1} \oplus P_i$. Then the algorithm $\mathscr{A}_7$ returns the set $Par(Q_t) = Par(P_1 \oplus \cdots \oplus P_t)$.*

Proposition 16.2 [1] *Let $P_1, \ldots, P_t$ be nonempty finite subsets of $\mathbb{R}^2$, $Q_1 = P_1$, and, for $i = 2, \ldots, t$, $Q_i = Q_{i-1} \oplus P_i$. Let $P_i^{(1)} = \{a : (a,b) \in P_i\}$ for $i = 1, \ldots, t$, m be a natural number, and $P_i^{(1)} \subseteq \{0, -1, \ldots, -m\}$ for $i = 1, \ldots, t$. Then, during the construction of the set $Par(Q_t)$, the algorithm $\mathscr{A}_7$ makes $O(t^2 m^2 \log(tm))$ additions and comparisons.*

16.2 Bi-Criteria Optimization of Matchings

We consider the problem of minimization of the total weight of edges in matchings instead of the problem of maximization of total weight. To this end, instead of positive integers (weights), we attach to edges additive inverse of weights.

Let G be a tree with the root r. A weight function w for G assigns a negative integer $w(e)$ to each edge e of the tree G. Let a be a matching in G. The value $w(a) = \sum_{e \in a} w(e)$ is called the weight of matching a relative to the weight function w. If $a = \lambda = \emptyset$, then $w(a) = 0$. Let w_1 and w_2 be weight functions for G. For any nonempty set A of matchings in G, we denote $P^{w_1,w_2}(A) = \{(w_1(a), w_2(a)) : a \in A\}$.

Lemma 16.1 *Let G be a tree, w_1 and w_2 be weight functions for G, and $A_1, \ldots, A_k$ be pairwise strongly disjoint nonempty sets of matchings in G. Then*

$$P^{w_1,w_2}(A_1 \otimes \cdots \otimes A_k) = P^{w_1,w_2}(A_1) \oplus \cdots \oplus P^{w_1,w_2}(A_k) .$$

Proof Let $a_1 \in A_1, \ldots, a_k \in A_k$. Since $A_1, \ldots, A_k$ are pairwise strongly disjoint, $w_1(a_1 \cup \cdots \cup a_k) = w_1(a_1) + \cdots + w_1(a_k)$ and $w_2(a_1 \cup \cdots \cup a_k) = w_2(a_1) + \cdots + w_2(a_k)$. Using these facts, it is easy to show that $P^{w_1,w_2}(A_1 \otimes \cdots \otimes A_k) = P^{w_1,w_2}(A_1) \oplus \cdots \oplus P^{w_1,w_2}(A_k)$. □

We corresponded a set $M_{D(G)}(u)$ of matchings in G to each node u of the graph $D(G)$ (see (14.1)–(14.4) for the case $\Delta = D(G)$). We correspond to each node u

of $D(G)$ the set of pairs of nonpositive integers $P_{D(G)}^{w_1,w_2}(u) = \{(w_1(a), w_2(a)) : a \in M_{D(G)}(u)\}$.

Let v be a terminal node of G. Then, using (14.1) we obtain

$$P_{D(G)}^{w_1,w_2}(v) = P_{D(G)}^{w_1,w_2}(v(\emptyset)) = \{(0,0)\} . \tag{16.1}$$

Let v be a nonterminal node of G, which has k leaving edges $e_1, \ldots, e_k$ entering nodes $v_1, \ldots, v_k$, respectively.

Using Lemmas 14.1 and 16.1 and (14.2)–(14.4), we obtain that, for $i = 1, \ldots, k$,

$$P_{D(G)}^{w_1,w_2}(v(\emptyset)) = \bigoplus_{j \in \{1,\ldots,k\}} P_{D(G)}^{w_1,w_2}(v_j) , \tag{16.2}$$

$$P_{D(G)}^{w_1,w_2}(v(e_i)) = \left(\bigoplus_{j \in \{1,\ldots,k\} \setminus \{i\}} P_{D(G)}^{w_1,w_2}(v_j) \right) \oplus P_{D(G)}^{w_1,w_2}(v_i(\emptyset)) \oplus \{(w_1(e_i), w_2(e_i))\} , \tag{16.3}$$

$$P_{D(G)}^{w_1,w_2}(v) = \bigcup_{\sigma \in \{e_1,\ldots,e_k,\emptyset\}} P_{D(G)}^{w_1,w_2}(v(\sigma)) . \tag{16.4}$$

Let G contain n nodes. We denote $E(G)$ the set of edges of G and $m = nW$, where $W = \max\{|w_1(e)| : e \in E(G)\}$. It is clear that, for any matching a in G,

$$w_1(a) \in \{0, -1, \ldots, -m\} . \tag{16.5}$$

We now describe the algorithm $\mathcal{A}_8$ for bi-criteria optimization of matchings in G, which constructs the set $Par(P_{D(G)}^{w_1,w_2}(r))$. In fact, this algorithm constructs, for each node u of $D(G)$, the set $B(u) = Par(P_{D(G)}^{w_1,w_2}(u))$ of Pareto optimal points for the set of pairs $P_{D(G)}^{w_1,w_2}(u)$.

Algorithm $\mathcal{A}_8$ (construction of POPs for matchings)

Input: A tree G with the root r, the graph $D(G)$, and weight functions w_1 and w_2 for G.

Output: The set $Par(P_{D(G)}^{w_1,w_2}(r))$ of Pareto optimal points for the set of pairs $P_{D(G)}^{w_1,w_2}(r) = \{(w_1(a), w_2(a)) : a \in M_{D(G)}(r)\}$.

1. If all nodes in G are processed, then return the set $B(r)$. Otherwise, choose a node v in the tree G, which is not processed yet and which is either a terminal node of G or a nonterminal node of G such that all children of v in G are processed.
2. If v is a terminal node of G, then set $B(v) = B(v(\emptyset)) = \{(0,0)\}$. Mark the node v as processed and proceed to step 1.
3. Let v be a nonterminal node of G, which has k leaving edges $e_1, \ldots, e_k$ entering nodes $v_1, \ldots, v_k$, respectively. Apply the algorithm $\mathcal{A}_7$ to sets $B(v_1), \ldots, B(v_k)$. Set $B(v(\emptyset))$ the output of the algorithm $\mathcal{A}_7$. Then, for $i = 1, \ldots, k$, apply the algorithm $\mathcal{A}_7$ to the sets

$$B(v_1), \ldots, B(v_{i-1}), B(v_i(\emptyset)), B(v_{i+1}), \ldots, B(v_k), \{(w_1(e_i), w_2(e_i))\}.$$

Set $B(v(e_i))$ the output of the algorithm $\mathscr{A}_7$. Construct the multiset $A(v) = \bigcup_{\sigma \in \{e_1,\ldots,e_k,\emptyset\}} B(v(\sigma))$ by simple transcription of elements from sets $B(v(\sigma))$, $\sigma \in \{e_1, \ldots, e_k, \emptyset\}$. Apply to the obtained multiset $A(v)$ the algorithm $\mathscr{A}_3$, which constructs the set $Par(A(v))$. Set $B(v) = Par(A(v))$. Mark the node v as processed and proceed to step 1.

Theorem 16.1 *Let G be a tree with the root r, n nodes, and the set of edges $E(G)$, and w_1, w_2 be weight functions for G. Then, for each node u of the graph $D(G)$, the algorithm $\mathscr{A}_8$ constructs the set $B(u) = Par(P^{w_1,w_2}_{D(G)}(u))$. To construct the set $B(r) = Par(P^{w_1,w_2}_{D(G)}(r))$, the algorithm $\mathscr{A}_8$ makes $O(n^6 W^2 \log(nW))$ comparisons and additions, where $W = \max\{|w_1(e)| : e \in E(G)\}$.*

Proof We will prove that $B(u) = Par(P^{w_1,w_2}_{D(G)}(u))$ for any node u of $D(G)$ by induction on nodes of G. At the same time, we will evaluate the number of arithmetical operations made by the algorithm $\mathscr{A}_8$.

Let v be a terminal node of G. From (16.1) it follows that

$$Par(P^{w_1,w_2}_{D(G)}(v)) = Par(P^{w_1,w_2}_{D(G)}(v(\emptyset))) = \{(0, 0)\}\ .$$

Therefore, $B(u) = Par(P^{w_1,w_2}_{D(G)}(u))$ for any $u \in \{v, v(\emptyset)\}$. We do not need arithmetical operations to construct the sets of Pareto optimal points for the sets of pairs corresponding to v and $v(\emptyset)$.

Let v be a nonterminal node of G, which has k leaving edges $e_1, \ldots, e_k$ entering nodes $v_1, \ldots, v_k$, respectively, and $B(u) = Par(P^{w_1,w_2}_{D(G)}(u))$ for any

$$u \in \{v_1, \ldots, v_k, v_1(\emptyset), \ldots, v_k(\emptyset)\}\ .$$

We now show that $B(u) = Par(P^{w_1,w_2}_{D(G)}(u))$ for any $u \in \{v, v(\emptyset), v(e_1) \ldots, v(e_k)\}$.

The equality $B(v(\emptyset)) = Par(P^{w_1,w_2}_{D(G)}(v(\emptyset))$ follows from the inductive hypothesis, equality (16.2) and Proposition 16.1. Using (16.5) and Proposition 16.2, we obtain that the construction of the set $Par(P^{w_1,w_2}_{D(G)}(v(\emptyset)))$ requires $O(k^2m^2 \log(km))$ additions and comparisons, where $m = nW$.

Let $i \in \{1, \ldots, k\}$. The equality $B(v(e_i)) = Par(P^{w_1,w_2}_{D(G)}(v(e_i))$ follows from the inductive hypothesis, equality (16.3), Proposition 16.1, and the fact that

$$Par(\{(w_1(e_i), w_2(e_i))\}) = \{(w_1(e_i), w_2(e_i))\}\ .$$

By (16.5) and Proposition 16.2, the construction of the set $Par(P^{w_1,w_2}_{D(G)}(v(e_i)))$ requires $O(k^2m^2 \log(km))$ additions and comparisons. To construct the set

$$Par(P^{w_1,w_2}_{D(G)}(v(e_i)))$$

for each $i \in \{1, \ldots, k\}$, we need $O(k^3m^2 \log(km))$ additions and comparisons.

From equality (16.4) and Lemma 4.3 it follows that

$$Par(P_{D(G)}^{w_1,w_2}(v)) \subseteq \bigcup_{\sigma \in \{e_1,\dots,e_k,\emptyset\}} Par(P_{D(G)}^{w_1,w_2}(v(\sigma))) .$$

By Lemma 4.2,

$$Par(P_{D(G)}^{w_1,w_2}(v)) = Par\left(\bigcup_{\sigma \in \{e_1,\dots,e_k,\emptyset\}} Par(P_{D(G)}^{w_1,w_2}(v(\sigma)))\right) .$$

The equality $B(v) = Par(P_{D(G)}^{w_1,w_2}(v))$ follows from this equality and the equalities $B(v(\sigma)) = Par(P_{D(G)}^{w_1,w_2}(v(\sigma)))$, $\sigma \in \{e_1, \dots, e_k, \emptyset\}$. We construct the set $A(v) = \bigcup_{\sigma \in \{e_1,\dots,e_k,\emptyset\}} Par(P_{D(G)}^{w_1,w_2}(v(\sigma)))$, which cardinality is, by (16.5) and Lemma 4.5, at most $(k+1)(m+1)$, and apply to it the algorithm $\mathscr{A}_3$. As a result, we obtain the set $Par(P_{D(G)}^{w_1,w_2}(v))$ using $O(km\log(km))$ comparisons—see Proposition 4.3. We proved that $B(u) = Par(P_{D(G)}^{w_1,w_2}(u))$ for any $u \in \{v, v(\emptyset), v(e_1)\dots, v(e_k)\}$. Therefore, $B(u) = Par(P_{D(G)}^{w_1,w_2}(u))$ for any node u of the graph $D(G)$.

To process a nonterminal node v of G with k children and all auxiliary nodes of $D(G)$ corresponding to v, we make $O(k^3m^2\log(km))$ comparisons and additions. The number of nonterminal nodes is at most n. It is clear that $k \le n$. Therefore, to construct the sets of POPs corresponding to all nodes of $D(G)$ we make $O(n^4m^2\log(nm))$ comparisons and additions, where $m = nW$. Then the algorithm $\mathscr{A}_5$ makes $O(n^6W^2\log(nW))$ comparisons and additions. □

16.3 Totally Optimal Matchings

Let G be a tree with the root r and w be a weight function for G that assigns to each edge in G a positive integer (weight). We denote by $w(G)$ the maximum total weight of edges in a matching in G, i.e., $w(G) = w^{opt}(M_{D(G)}(r))$. Let w_1 and w_2 be weight functions for G and a be a matching in G. We will say that the matching a is totally optimal relative to w_1 and w_2 if $w_1(a) = w_1(G)$ and $w_2(a) = w_2(G)$.

There are two ways to recognize experimentally the existence of a totally optimal matching in G relative to w_1 and w_2.

The first one is based on the use of a multi-stage optimization technique. We construct the graph $D(G)$. For $i = 1, 2$, we apply to $D(G)$ the procedure of matching optimization (algorithm $\mathscr{A}_6$) relative to w_i. As a result, we obtain the graphs $D(G)^{w_1}$ and $D(G)^{w_2}$, and numbers $w_1(G)$ and $w_2(G)$ attached to the node r of these graphs, respectively (see Theorem 15.2). Next, we apply to $D(G)$ sequentially the procedures of matching optimization relative to w_1 and w_2. As a result, we obtain the graph $D(G)^{w_1,w_2}$ and a number $g(r)$ attached to the node r of this graph. From Theorem 15.2 it follows that $g(r) = \max\{w_2(a) : a \in Opt_{w_1}(M_{D(G)}(r))\}$, i.e., $g(r)$ is the maximum w_2-weight of a matching among all matchings with the maximum

w_1-weight (here w_i-weight of a matching a is $w_i(a)$). One can show that a totally optimal matching in G relative to w_1 and w_2 exists if and only if $g(r) = w_2(G)$.

The second way is to construct the set $Par(P_{D(G)}^{w_1,w_2}(r))$ of all POPs for the bi-criteria optimization problem for matchings in G relative to w_1 and w_2. A totally optimal matching in G relative to w_1 and w_2 exists if and only if the set $Par(P_{D(G)}^{w_1,w_2}(r))$ contains only one POP.

There are some theoretical results regarding totally optimal matchings relative to c and w, where c is the weight function, which assigns to each edge the weight 1 and w is a weight function satisfying some special conditions. Note that $c(G)$ is the maximum cardinality of a matching in G.

Let κ be a function that assigns positive integers (weights) to nodes of the tree G. We correspond to κ a weight function w_κ, which assigns to each edge $e = (u, v)$ of the tree G the weight $w_\kappa(e) = \kappa(u) + \kappa(v)$. Let a be a matching in G. It is known [2, 3, 5] that if $w_\kappa(a) = w_\kappa(G)$, then $c(a) = c(G)$, i.e., each matching with the maximum w_κ-weight has the maximum cardinality. It means that there exists a totally optimal matching in G relative to w_κ and c.

Let G have n nodes, $E(G)$ be the set of edges in G, and w be a weight function for G. We denote $h = \min\{w(e) : e \in E(G)\}$ and $H = \max\{w(e) : e \in E(G)\}$. Let $H > h$ and $\frac{H}{H-h} > \lfloor \frac{n}{2} \rfloor$. In this case, for each matching a in G, if $w(a) = w(G)$, then $c(a) = c(G)$, i.e., each matching with the maximum w-weight has the maximum cardinality [2]. It means that there exists a totally optimal matching in G relative to w and c.

In the next chapter, we study experimentally the existence of totally optimal matchings relative to c and a randomly generated weight function w.

References

1. AbouEisha, H., Amin, T., Chikalov, I., Hussain, S., Moshkov, M.: Extensions of Dynamic Programming for Combinatorial Optimization and Data Mining, Intelligent Systems Reference Library, vol. 146. Springer (2019)
2. Gentry, S., Mankowski, M., Michael, T.: Maximum matchings in graphs for allocating kidney paired donation. Oper. Res. Health Care **25** (2020)
3. Halappanavar, M.: Algorithms for vertex-weighted matching in graphs. Ph.D. thesis, Old Dominion University, Norfolk, VA, USA (2009)
4. Mankowski, M., Moshkov, M.: Bi-criteria optimization of matchings in trees with application to kidney exchange. In: 26th International Workshop on Concurrency, Specification and Programming, CS&P 2017, Warsaw, Poland, September 25–27, 2017 (2017). http://csp2017.mimuw.edu.pl/data/uploads/papers/CSP2017_paper_1.pdf
5. Roth, A., Sönmez, T., Unver, U.: Pairwise kidney exchange. J. Econom. Theory **125**(2), 151–188 (2005)

Chapter 17
Example and Experimental Results

In this chapter, we consider an example of multi-stage and bi-criteria optimization of matchings for a tree with five nodes [1, 2]. We also discuss the results of computational experiments with the algorithms $\mathscr{A}_5$, $\mathscr{A}_6$, and $\mathscr{A}_8$ for randomly generated trees and weight functions. We compare the number of matchings before and after optimization, compare the maximum cardinality of matchings before and after optimization relative to a randomly generated weight function, consider the number of POPs for bi-criteria optimization of matchings relative to the cardinality and a weight function, and study the existence of totally optimal matchings.

17.1 Example

We consider an example of multi-stage and bi-criteria optimization of matchings for a tree G with the root v_1 depicted in Fig. 17.1a, which has five nodes v_1, v_2, v_3, v_4, v_5 and four edges e_1, e_2, e_3, e_4.

We study two weight functions. The first weight function w_1 corresponds the weight 1 to each edge (we maximize the cardinality of matching). The second weight function w_2 corresponds weights $2, 5, 1, 2$ to edges e_1, e_2, e_3, e_4, respectively (we maximize the weight of matching relative to this function)—see Fig. 17.1a.

There are two matchings with the maximum cardinality $\{e_1, e_3\}$ and $\{e_1, e_4\}$, and one matching $\{e_2\}$ with the maximum total edge weight relative to w_2.

Our first aim is to find all matchings with the maximum total edge weight among all matchings with the maximum cardinality.

We construct the graph $D(G)$ (see Fig. 17.1b) and apply to it the procedure of optimization (algorithm $\mathscr{A}_6$) relative to the weight function w_1. As a result, we obtain the edge-preserving subgraph $D(G)^{w_1}$ of the graph $D(G)$ (see Fig. 17.2a). It is easy to see that $M_{D(G)^{w_1}}(v_1) = \{\{e_1, e_3\}, \{e_1, e_4\}\}$. This is exactly the set of all matchings in G with the maximum cardinality.

M. Mankowski and M. Moshkov, *Dynamic Programming Multi-Objective Combinatorial Optimization*, Studies in Systems, Decision and Control 331,
https://doi.org/10.1007/978-3-030-63920-4_17

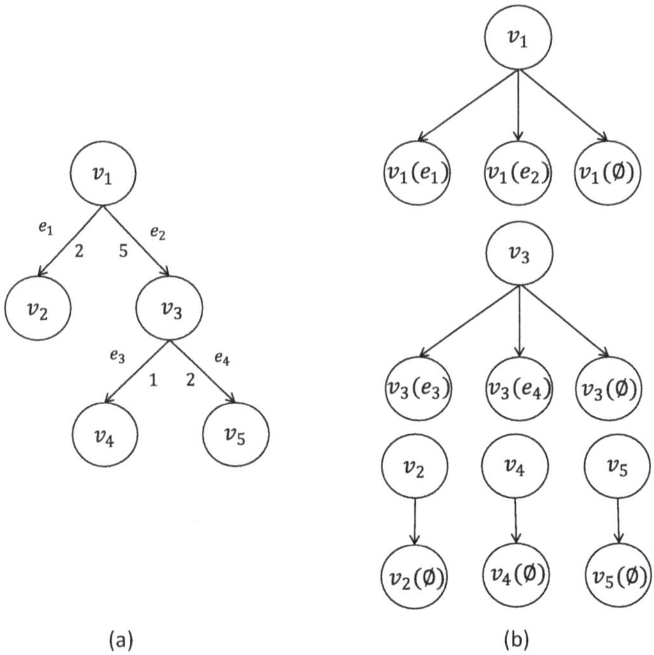

Fig. 17.1 Graph $D(G)$ (**b**) corresponding to tree G (**a**)

We apply to the graph $D(G)^{w_1}$ the procedure of optimization (algorithm $\mathscr{A}_6$) relative to the weight function w_2.

As a result, we obtain the edge-preserving subgraph $D(G)^{w_1,w_2}$ of the graph $D(G)$ (see Fig. 17.2b). The set $M_{D(G)^{w_1,w_2}}(v_1) = \{\{e_1, e_4\}\}$ contains the only matching with the maximum total edge weight among all matchings with the maximum cardinality.

Our second aim is to construct all POPs for the problem of bi-criteria optimization of matchings relative to the weight functions w_1 and w_2.

We apply to the graph $D(G)$ the procedure of bi-criteria optimization (algorithm $\mathscr{A}_8$) relative to the weight functions w_1 and w_2. As a result, each node of $D(G)$ is labeled with the set of POPs corresponding to the set of matchings associated with this node (see Fig. 17.3). In particular, the whole set of matchings (it is associated with the node v_1) has two POPs $(1, 5)$ and $(2, 4)$.

17.2 Experimental Results

To evaluate created methods and to clarify some aspects of the matching optimization problems, we conduct computational experiments with randomly generated rooted trees, with randomly generated weight function w and with the weight function c,

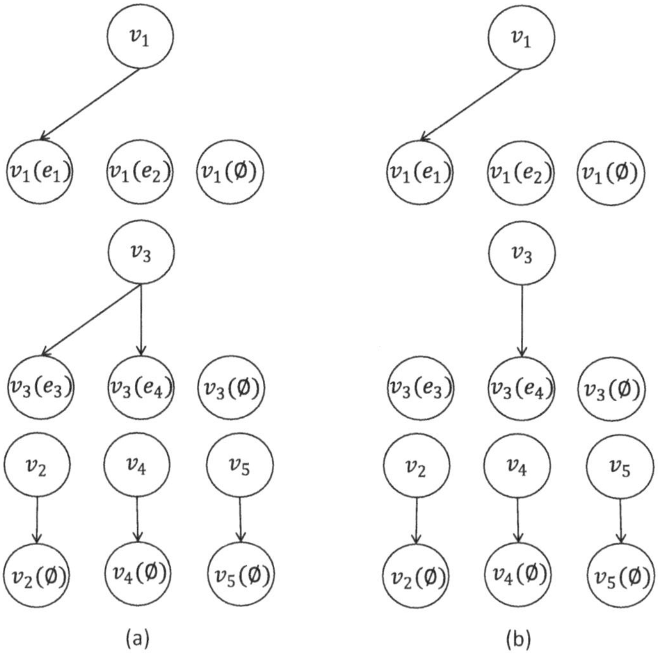

Fig. 17.2 Graph $D(G)^{w_1}$ (**a**) and graph $D(G)^{w_1,w_2}$ (**b**)

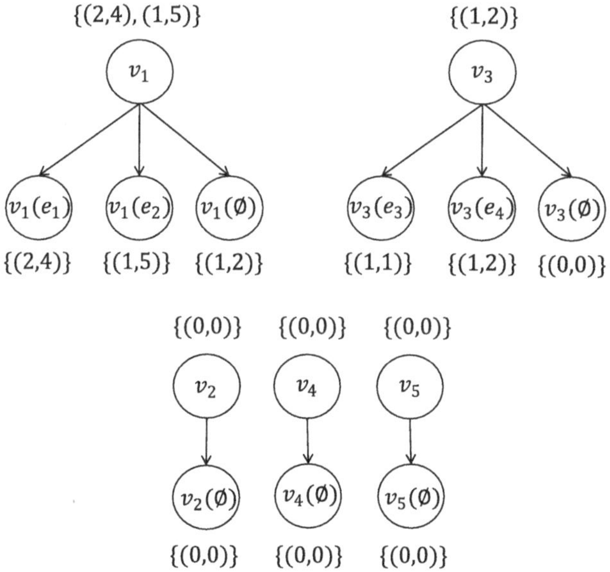

Fig. 17.3 Sets of POPs attached to nodes of $D(G)$

which assigns the weight 1 to each edge of the tree. We repeat each experiment 100 times. For each trial, we construct a tree and a weight function w for it using the following random procedure. First, we define the range of a possible number of children "# children" and the range of a possible edge weight "weights". Both ranges are given in the form 1-m, where m is a natural number.

For the root node, we generate a random number of children in the given range "# children", add edges connecting the root node with the children, and assign weights randomly in the given range "weights" to these edges. Then from all terminal nodes of the constructed tree, we randomly choose one and add children to this node, etc. The procedure terminates when the total number of nodes in the constructed tree is equal to the given limit "# nodes". In all random processes, we use uniform distributions.

In Tables 17.1 and 17.2, we show the average number of matchings in the initial tree (column "∅"), after optimization relative to c (column "c"), after optimization relative to w (column "w"), after optimization relative to c and w (column "c, w"), and after optimization relative to w and c (column "w, c") among 100 trials. In Table 17.1, we consider ranges # children 1–2 and weights 1–10, and in Table 17.2—ranges # children 1–15 and weights 1–100.

We see that the initial number of matchings and the number of matchings with the maximum cardinality are large, but the number of matchings after optimization relative to w is relatively small. We also see the following trends: in Table 17.2 in comparison with Table 17.1, the initial number of matchings in trees is decreasing, the number of matchings after optimization relative to c is increasing, and the number of matchings after any sequence of optimizations containing w is decreasing.

Table 17.1 Average number of matchings in the initial tree (∅), after optimization relative to c (c), after optimization relative to w (w), after optimization relative to c and w (c, w), and after optimization relative to w and c (w, c) among 100 trials; # children 1–2; weights 1–10

# nodes	Average number of matchings				
	∅	c	w	c, w	w, c
10	7.1×10^1	4.7	1.4	1.2	1.2
30	6.1×10^5	1.4×10^2	2.1	1.6	1.8
50	5.1×10^9	2.3×10^3	4.0	2.1	2.4
70	4.2×10^{13}	7.6×10^4	5.6	3.4	4.1
90	3.6×10^{17}	7.8×10^5	7.0	3.4	4.0
110	2.8×10^{21}	1.2×10^7	1.5×10^1	6.6	8.0
130	2.7×10^{25}	4.1×10^8	1.7×10^1	6.4	8.0
150	2.3×10^{29}	1.7×10^{10}	3.2×10^1	6.7	10.1
170	1.8×10^{33}	7.5×10^{10}	5.6×10^1	10.6	15.6
190	1.6×10^{37}	1.6×10^{12}	8.7×10^1	14.8	23.2
210	1.4×10^{41}	2.3×10^{13}	2.0×10^2	19.2	46.8
230	1.1×10^{45}	1.2×10^{14}	2.4×10^2	28.2	48.5

Table 17.2 Average number of matchings in the initial tree (∅), after optimization relative to c (c), after optimization relative to w (w), after optimization relative to c and w (c, w), and after optimization relative to w and c (w, c) among 100 trials; # children 1–15; weights 1–100

# nodes	Average number of matchings				
	∅	c	w	c, w	w, c
10	3.1×10^1	7.8	1.0	1.0	1.0
30	5.3×10^3	6.6×10^2	1.2	1.2	1.2
50	1.1×10^6	6.1×10^4	1.2	1.2	1.2
70	1.2×10^8	4.8×10^6	1.3	1.3	1.3
90	2.4×10^{10}	4.0×10^8	1.6	1.6	1.6
110	4.1×10^{12}	3.2×10^{10}	1.5	1.5	1.5
130	4.8×10^{14}	2.8×10^{12}	1.7	1.7	1.7
150	7.7×10^{16}	1.9×10^{14}	1.6	1.6	1.6
170	1.8×10^{19}	1.3×10^{16}	2.0	2.0	2.0
190	1.5×10^{22}	1.5×10^{18}	2.0	2.0	2.0
210	2.7×10^{23}	1.0×10^{20}	2.3	2.3	2.3
230	3.4×10^{26}	1.0×10^{22}	2.9	2.9	2.9

In Table 17.3, we compare the maximum cardinality of matchings before and after optimization relative to w among 100 trials. In this table, we use format ${}_{min}average_{max}$. We see that the optimization relative to w, as the first step, will not lead to a considerable decrease in the average of the maximum cardinality of matchings. With the increase in the number of children and weight ranges, the maximum cardinality of matchings before and after optimization relative to w is decreasing on average.

In Table 17.4, we consider the number of Pareto optimal points (POPs) (in format ${}_{min}average_{max}$) for problem of bi-criteria optimization of matchings relative to c and w with weights 1–10 among 100 trials. The most interesting result of these experiments is the small number of POPs—at most 12 (Fig. 17.4 shows all nine POPs for one of the considered problems). The minimum number of POPs in each series of 100 trials except one case is one. It means that in these series, there are trees with totally optimal matchings relative to c and w. In two series, the maximum number of POPs is equal to one. It means that each tree from these series has a totally optimal matching relative to c and w.

In Table 17.5, we show the number of trees with totally optimal matchings relative to c and w with weights 1–10 among 100 trials. In many series, the number of trees with totally optimal matchings is close to 100.

Table 17.3 Maximum cardinality of matchings (in format ${}_{min}average_{max}$) before and after optimization relative to w among 100 trials

# nodes	# children	Weights	Maximum matching cardinality	
			before optimization	after optimization
10	1–2	1–10	${}_{4}4.37_{5}$	${}_{3}4.12_{5}$
30	1–2	1–10	${}_{11}13.22_{15}$	${}_{11}12.57_{14}$
50	1–2	1–10	${}_{20}22.02_{24}$	${}_{18}20.78_{23}$
70	1–2	1–10	${}_{28}30.82_{33}$	${}_{26}28.96_{31}$
90	1–2	1–10	${}_{36}39.97_{43}$	${}_{32}37.77_{42}$
110	1–2	1–10	${}_{46}48.35_{52}$	${}_{42}45.85_{49}$
130	1–2	1–10	${}_{55}57.74_{61}$	${}_{50}54.62_{59}$
150	1–2	1–10	${}_{62}66.33_{71}$	${}_{57}62.69_{67}$
170	1–2	1–10	${}_{72}75.69_{81}$	${}_{65}71.33_{76}$
190	1–2	1–10	${}_{79}84.40_{90}$	${}_{74}79.71_{86}$
210	1–2	1–10	${}_{89}93.33_{98}$	${}_{82}87.98_{93}$
230	1–2	1–10	${}_{98}102.22_{106}$	${}_{90}96.40_{101}$
10	1–15	1–100	${}_{1}2.64_{5}$	${}_{1}2.61_{5}$
30	1–15	1–100	${}_{2}5.13_{8}$	${}_{2}5.12_{8}$
50	1–15	1–100	${}_{4}7.59_{13}$	${}_{4}7.52_{13}$
70	1–15	1–100	${}_{7}9.53_{14}$	${}_{7}9.47_{14}$
90	1–15	1–100	${}_{8}12.29_{18}$	${}_{8}12.20_{18}$
110	1–15	1–100	${}_{11}14.84_{20}$	${}_{11}14.75_{20}$
130	1–15	1–100	${}_{12}17.33_{23}$	${}_{12}17.27_{23}$
150	1–15	1–100	${}_{15}19.66_{26}$	${}_{15}19.64_{26}$
170	1–15	1–100	${}_{17}22.40_{31}$	${}_{17}22.30_{31}$
190	1–15	1–100	${}_{18}25.02_{35}$	${}_{18}24.92_{35}$
210	1–15	1–100	${}_{19}27.12_{33}$	${}_{19}27.03_{33}$
230	1–15	1–100	${}_{22}30.22_{39}$	${}_{22}30.13_{39}$

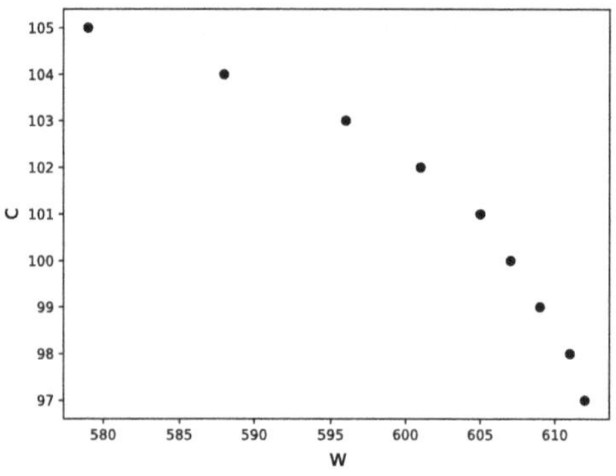

Fig. 17.4 Pareto optimal points for a bi-criteria optimization problem with parameters # nodes 230, # children 1–2, and weights 1–10

Table 17.4 Number of Pareto optimal points (in format ${}_{min}average_{max}$) for bi-criteria optimization problem relative to c and w with weights 1–10 among 100 trials

# children	# nodes					
	30	70	110	150	190	230
1–2	${}_{1}1.72_{3}$	${}_{1}2.60_{5}$	${}_{1}3.49_{7}$	${}_{1}4.71_{10}$	${}_{1}5.50_{12}$	${}_{2}6.22_{11}$
1–4	${}_{1}1.22_{2}$	${}_{1}1.82_{6}$	${}_{1}2.24_{5}$	${}_{1}2.52_{6}$	${}_{1}2.92_{6}$	${}_{1}3.56_{8}$
1–6	${}_{1}1.16_{3}$	${}_{1}1.44_{3}$	${}_{1}1.54_{4}$	${}_{1}1.70_{4}$	${}_{1}1.89_{4}$	${}_{1}1.92_{7}$
1–8	${}_{1}1.04_{2}$	${}_{1}1.12_{3}$	${}_{1}1.23_{4}$	${}_{1}1.24_{4}$	${}_{1}1.41_{4}$	${}_{1}1.40_{4}$
1–10	${}_{1}1.02_{2}$	${}_{1}1.08_{2}$	${}_{1}1.14_{3}$	${}_{1}1.12_{2}$	${}_{1}1.25_{4}$	${}_{1}1.18_{3}$
1–12	${}_{1}1.02_{2}$	${}_{1}1.06_{2}$	${}_{1}1.10_{3}$	${}_{1}1.06_{2}$	${}_{1}1.11_{2}$	${}_{1}1.07_{3}$
1–14	${}_{1}1.02_{2}$	${}_{1}1.04_{2}$	${}_{1}1.03_{2}$	${}_{1}1.03_{2}$	${}_{1}1.04_{2}$	${}_{1}1.02_{2}$
1–16	${}_{1}1.04_{2}$	${}_{1}1.02_{2}$	${}_{1}1.01_{2}$	${}_{1}1.07_{2}$	${}_{1}1.04_{2}$	${}_{1}1.03_{2}$
1–18	${}_{1}1.01_{2}$	${}_{1}1.02_{2}$	${}_{1}1.00_{1}$	${}_{1}1.00_{1}$	${}_{1}1.03_{2}$	${}_{1}1.03_{2}$
1–20	${}_{1}1.03_{2}$	${}_{1}1.01_{2}$	${}_{1}1.01_{2}$	${}_{1}1.01_{2}$	${}_{1}1.02_{2}$	${}_{1}1.01_{2}$

Table 17.5 Number of trees with totally optimal matchings relative to c and w with weights 1–10 among 100 trials

# children	# nodes					
	30	70	110	150	190	230
1–2	39	14	2	1	1	0
1–4	78	42	29	20	13	4
1–6	86	60	57	50	39	36
1–8	96	89	81	78	66	69
1–10	98	92	87	88	82	85
1–12	98	94	91	94	89	94
1–14	98	96	97	97	96	98
1–16	96	98	99	93	96	97
1–18	99	98	100	100	97	97
1–20	97	99	99	99	98	99

References

1. Mankowski, M., Moshkov, M.: Bi-criteria optimization of matchings in trees with application to kidney exchange. In: 26th International Workshop on Concurrency, Specification and Programming, CS&P 2017, Warsaw, Poland, September 25–27, 2017 (2017). http://csp2017.mimuw.edu.pl/data/uploads/papers/CSP2017_paper_1.pdf
2. Mankowski, M., Moshkov, M.: Multi-stage optimization of matchings in trees with application to kidney exchange. In: Polkowski, L., Yao, Y., Artiemjew, P., Ciucci, D., Liu, D., Slezak, D., Zielosko, B. (eds.) Proceedings of the International Joint Conference on Rough Sets (IJCRS 2017), Part I, Olsztyn, Poland, July 3–7, 2017. Lecture Notes in Computer Science, vol. 10313, pp. 123–130. Springer (2017)

Part IV
Combinatorial Optimization Problems. Syntactical Circuits

In this part, we study two combinatorial optimization problems for which we cannot find conventional circuits without repetitions: optimization of matchings in trees and 0/1 knapsack problem. For each problem, we construct a conventional circuit with repetitions, transform this circuit into a syntactical circuit without repetitions, and define at least two increasing or strictly increasing cost functions for the syntactical circuit each of which is adapted (see Sect. 2.4). We evaluate the number of operations and the time required by the algorithms $\mathcal{A}_1$, $\mathcal{A}_2$, and $\mathcal{A}_4$, consider an example of the problem instance, and discuss the results of experiments with randomly generated instances of the problem. For the considered problems and cost functions, the algorithms $\mathcal{A}_1$, $\mathcal{A}_2$, and $\mathcal{A}_4$ work with the schemes of circuits instead of the circuits (see Sect. 2.4).

In Chap. 18, we study the problem of optimization of matchings in trees.

In Chap. 19, we consider the 0/1 knapsack problem.

Chapter 18
Optimization of Matchings in Trees

We consider the problem of optimization of matchings in trees studied in Part III, where separate algorithms for optimization and counting of matchings have been introduced based on the results published in [2, 3].

In this chapter, we design a syntactical circuit for the matching optimization problem, describe cost functions, evaluate the number of operations and the time required by the algorithms for optimization and counting, discuss an example, show experimental results, and compare results obtained in this chapter and Part III.

18.1 Definition of Circuit and Cost Functions

Let G be a tree with $n \geq 1$ nodes and $n - 1$ edges. We choose a node r in the tree G as the root. It will be useful for us to consider G as a directed graph with the orientation of edges from the root r. The tree G is given by its adjacency list. A matching in G is a set of edges without common nodes.

The number n is the dimension of the problems under consideration related to the optimization of matchings in trees. To define a cost function for a matching optimization problem, we need to give $n - 1$ positive integer weights of edges. To describe an instance of the problem, we need the adjacency list and one or two cost functions. Note that n is at most the number of numeric parameters of the problem plus one.

The adjacency list and the binary representations of $n - 1$ or $2n - 2$ numbers defining one or two cost functions form the representation of the problem instance (we will omit the word instance). It is clear that n is at most the length of the problem representation.

Each node y in G defines a subtree $G(y)$ of G in which y is the root. We denote by $M_G(y)$ the set of matchings in $G(y)$ and by $M_G(y(\emptyset))$ we denote the set of matchings in $G(y)$, which do not contain edges that are incident to the node y. Let C and D be

M. Mankowski and M. Moshkov, *Dynamic Programming Multi-Objective Combinatorial Optimization*, Studies in Systems, Decision and Control 331,
https://doi.org/10.1007/978-3-030-63920-4_18

nonempty sets of matchings in G. We denote $C \otimes D = \{c \cup d : c \in C, d \in D\}$. We now describe the set $M_G(r)$. Let y be a terminal node of G. Then

$$M_G(y) = M_G(y(\emptyset)) = \{\lambda\} ,$$

where λ is the empty matching.

Let y be a nonterminal node of G, which has k leaving edges $e_1, \ldots, e_k$ entering nodes $y_1, \ldots, y_k$, respectively. Then, for $i = 1, \ldots, k$,

$$M_G(y(\emptyset)) = \bigotimes_{j \in \{1,\ldots,k\}} M_G(y_j) ,$$

$$M_G(y(e_i)) = \left(\bigotimes_{j \in \{1,\ldots,k\} \setminus \{i\}} M_G(y_j) \right) \otimes M_G(y_i(\emptyset)) \otimes \{\{e_i\}\} ,$$

$$M_G(y) = \bigcup_{\sigma \in \{e_1,\ldots,e_k,\emptyset\}} M_G(y(\sigma)) ,$$

where $M_G(y(e_i))$ is the set of matchings in $G(y)$ containing the edge e_i.

Set $U = \bigcup_{y \in V(G)} M_G(y)$, where $V(G)$ is the set of all nodes of G. We now describe a circuit S over the set U, which represents the set $M_G(r)$.

For each terminal node y from G, the circuit S contains two input nodes $v(y)$ and $v(y, \emptyset)$ labeled with $\{\lambda_y\}$ and $\{\lambda_{y,\emptyset}\}$, respectively. Both, λ_y and $\lambda_{y,\emptyset}$, denote the empty matching λ. For each nonterminal node y from G, the circuit S contains one unifying node $v(y)$ labeled with the symbol $\cup$. The unifying node $v(r)$ is considered as the output and is labeled with $*$.

For each nonterminal node y from G with k leaving edges $e_1, \ldots, e_k$ entering nodes $y_1, \ldots, y_k$, the circuit S contains a functional node $v(y, e_i)$, $i = 1, \ldots, k$, labeled with the function

$$\mu_{y,e_i}(x_1, \ldots, x_k) = \left(\bigcup_{j \in \{1,\ldots,k\} \setminus \{i\}} x_j \right) \cup x_i \cup \{e_i\}$$

from $\prod_{j=1,\ldots,i-1} M_G(y_j) \times M_G(y_i(\emptyset)) \times \prod_{j=i+1,\ldots,k} M_G(y_j)$ to $M_G(y(e_i))$. Additionally, for the same nonterminal node y, the circuit S contains one functional node $v(y, \emptyset)$ labeled with the function $\mu_{y,\emptyset}(x_1, \ldots, x_k) = \bigcup_{j \in \{1,\ldots,k\}} x_j$ from

$$\prod_{j=1,\ldots,k} M_G(y_j)$$

to $M_G(y(\emptyset))$.

For each nonterminal node y from G with k leaving edges $e_1, \dots, e_k$ entering nodes $y_1, \dots, y_k$, the unifying node $v(y)$ has k entering edges leaving the nodes $v(y, e_i), i = 1, \dots, k$, and one entering edge from the node $v(y, \emptyset)$. For the considered node y, each functional node $v(y, e_i)$ has k entering edges leaving the nodes $v(y_1), \dots, v(y_{i-1}), v(y_i, \emptyset), v(y_{i+1}), \dots, v(y_k)$. We label these edges with the numbers $1, 2, \dots, k$ (we will not show these numbers in the example). For the considered node y, the functional node $v(y, \emptyset)$ has k entering edges leaving the unifying nodes $v(y_1), \dots, v(y_k)$. We label these edges with the numbers $1, 2, \dots, k$ (we will not show these numbers in the example).

One can show that, for any node $v(y) \in V_i(S) \cup V_u(S)$, $S(v(y)) = M_G(y)$. For any node $v(y, \emptyset) \in V_i(S) \cup V_f(S)$, $S(v(y, \emptyset)) = M_G(y(\emptyset))$ and, for any $v(y, e_i) \in V_f(S)$, $S(v(y, e_i)) = M_G(y(e_i))$.

The circuit S contains at most $2n$ input nodes, at most n unifying nodes with at most n entering edges each, and at most $2n$ functional nodes with at most n entering edges each. Hence, it contains at most $5n$ nodes and at most $3n^2$ edges.

The circuit S is not a circuit without repetitions: both input and functional nodes might share some of the elements. We correspond to S the syntactical circuit $\hat{S}$, which is a circuit without repetitions (see Sect. 2.3).

Let y be a terminal node of G. In the syntactical circuit $\hat{S}$, we use element symbols $\hat{\lambda}_y$ and $\hat{\lambda}_{y,\emptyset}$ instead of the elements (empty matchings) λ_y and $\lambda_{y,\emptyset}$. Let y be a nonterminal node of G, which has k leaving edges $e_1, \dots, e_k$. In the syntactical circuit $\hat{S}$, we use functional symbols (and functions) $\hat{\mu}_{y,e_i}, i = 1, 2, \dots, k$ and $\hat{\mu}_{y,\emptyset}$ instead of functions $\mu_{y,e_i}, i = 1, 2, \dots, k$, and $\mu_{y,\emptyset}$.

Let w be a weight function, which assigns positive integers (weights) to edges of the tree G. We now define a cost function Ψ^w for the circuit $\hat{S}$ given by the functions ψ^w and $\psi^w_v, v \in V_f(\hat{S})$. For any terminal node y of G, $\psi^w(\hat{\lambda}_{y,\emptyset}) = \psi^w(\hat{\lambda}_y) = 0$. Let y be a nonterminal node of G with k leaving edges $e_1, \dots, e_k$. Then $\psi^w_{v(y,\emptyset)}(x_1, \dots, x_k) = x_1 + \dots + x_k$ and $\psi^w_{v(y,e_i)}(x_1, \dots, x_k) = x_1 + \dots + x_k + (-w(e_i))$ for $i = 1, \dots, k$.

This cost function is adapted (see Sect. 2.4). It is described by formulas over the set of elementary operations $B = \{x + y\}$—we assume that already have value $-w(e)$ for each edge e of the tree G.

To compute ψ^w, we do not need any operations, while to compute $\psi^w_v, v \in V_f(\hat{S})$ we need at most $k(v)$ operations from B.

The function Ψ^w is a strictly increasing nonpositive integer cost function. Denote $f(x_1, x_2) = x_1 + x_2$. For $k = 2, 3, \dots$, we define the function $f_k(x_1, \dots, x_k)$ in the following way: $f_2(x_1, x_2) = f(x_1, x_2)$ and $f_k(x_1, \dots, x_k) = f(f_{k-1}(x_1, \dots, x_{k-1}), x_k)$ for $k \geq 3$. Then $x_1 + \dots + x_k = f_k(x_1, \dots, x_k)$ and

$$x_1 + \dots + x_k + (-w(e_i)) = f_{k+1}(x_1, \dots, x_k, -w(e_i)) .$$

Note that $f(x_1, x_2)$ is an increasing bounded nonpositive integer function. It means, we can use the modification of the algorithms $\mathcal{A}_4$ described in Sect. 4.3 to process functional nodes of $\hat{S}$.

Let δ be a formula from $\hat{S}(v^*)$. Then the element $a(\delta)$ from $S(v^*)$ represented by δ is a matching. It is easy to see that $\Psi^w(\delta) = -\Sigma_{e \in a(\delta)} w(e)$. Therefore, Ψ^w is a proper cost function for the circuit $\hat{S}$ relative to the circuit S.

18.2 Time Complexity of Algorithms

The algorithms $\mathscr{A}_1$ and $\mathscr{A}_2$, and the modification of the algorithm $\mathscr{A}_4$ work with the scheme of the circuit $\hat{S}$ (see Sect. 2.4). The circuit $\hat{S}$ is completely described by the tree G. Therefore, the time complexity of the scheme of the circuit $\hat{S}$ construction is linear depending on the number of its nodes and edges, i.e., $O(n^2)$.

Note that the algorithms $\mathscr{A}_1$ and $\mathscr{A}_2$ can also work with the schemes of edge-preserving subcircuits of $\hat{S}$ obtained as a result of the algorithm $\mathscr{A}_1$ work—see Remark 3.1.

Denote $B^+ = B \cup \{x \cdot y, x : y\}$, where $x : y$ is the operation of comparison with values $x < y$, $x = y$, and $x > y$.

Let w be a weight function for G. We denote $W(w) = \max\{w(e) : e \in E(G)\}$, where $E(G)$ is the set of edges of G. One can show that $N_{\Psi^w}(\hat{S}) \leq (n-1)W(w) \leq nW(w) - 1$, where $N_{\Psi^w}(\hat{S}) = \max\{|\Psi^w(\delta)| : \delta \in U(\hat{S})\}$ and $U(\hat{S}) = \bigcup_{v \in V(\hat{S})} \hat{S}(v)$. It is clear that $\log_2 W(w)$ is at most the length of the representation of a problem that uses the cost function Ψ^w.

We can use the algorithm $\mathscr{A}_1$ for multi-stage optimization of formulas (matchings). Using Theorem 3.1, we evaluate the number of operations from B^+ required by the algorithm $\mathscr{A}_1$ working with the scheme of an edge-preserving subcircuit T of the circuit $\hat{S}$ and a cost function Ψ^w for $\hat{S}$, where w is a weight function for G.

The scheme of T contains at most $2n$ input nodes. We do not need any operations to process these nodes.

The scheme of T contains at most n unifying nodes with at most n entering edges each. To treat these nodes, the algorithm $\mathscr{A}_1$ makes at most $(2n-1)n \leq 2n^2$ comparisons.

The scheme of T contains at most $2n$ functional nodes with at most n entering edges each. For each functional node v, the algorithm makes at most n operations from B (additions) to compute ψ_v^w. The total number of operations for all functional nodes is at most $2n^2$. The overall number of operations from B^+ (comparisons and additions) made by $\mathscr{A}_1$ is at most $2n^2 + 2n^2 \leq 4n^2$, i.e., it is polynomial depending on the dimension n.

In the software model of computation, for the considered scheme of the circuit and cost function, the algorithm $\mathscr{A}_1$ has polynomial time complexity $O(n^2)$ depending on the problem dimension n.

Using Remark 3.2 we obtain that, for the algorithm $\mathscr{A}_1$, the number $nW(w)$ is an upper bound on the absolute values of the numbers at the inputs of the operations ψ_v^w related to the functional nodes and at the inputs of comparisons related to the unifying nodes of S. It is easy to check (see Remark 3.2) that, for the algorithm $\mathscr{A}_1$, the number $nW(w)$ is an upper bound on the absolute values of the numbers at the inputs of operations from B in formulas for the functions ψ_v^w. The algorithm uses only the operations of addition and comparison. Therefore, in the integer model of computation, for the considered scheme of the circuit and the cost function, the algorithm $\mathcal{A}_1$ has the time complexity $O(n^2 \log(nW(w)))$, which is polynomial depending on the length of the problem representation (n and $\log_2 W(w)$ are at most the length of the problem representation).

To count all formulas described by T, we apply the algorithm $\mathcal{A}_2$. We use Theorem 3.2 to evaluate the number of operations required by $\mathcal{A}_2$. The scheme of T contains at most n unifying nodes with at most n entering edges each. Therefore, we need at most n^2 additions to treat these unifying nodes. For all of at most $2n$ functional nodes with at most n entering edges each, we need at most $2n^2$ operations of multiplication. To treat the input nodes, we do not need any operations. Thus, the total number of operations from B^+ made by $\mathcal{A}_2$ is at most $3n^2$.

In the software model of computation, for the considered scheme of the circuit, the algorithm $\mathcal{A}_2$ has polynomial time complexity $O(n^2)$ depending on the problem dimension n.

We will not consider the time complexity of the algorithm $\mathcal{A}_2$ in the integer model of computation since we cannot find reasonable upper bounds on the number of formulas described by the circuit $\hat{S}$.

Let w_1 and w_2 be two weight functions for G. We consider two cost functions ψ^{w_1} and ψ^{w_2} for the circuit $\hat{S}$. We can use the modified algorithm $\mathcal{A}_4$ (see Sect. 4.3) to find the set of Pareto optimal points for the problem of bi-criteria optimization of formulas (matchings) relative to ψ^{w_1} and ψ^{w_2}.

Let $M = M_{\psi^{w_1},\psi^{w_2}}(\hat{S}) = \max\{|Par(t_{\psi^{w_1},\psi^{w_2}}(\hat{S}, v))| : v \in V(\hat{S})\}$. By Proposition 4.4, $M \leq N_{\psi^{w_1}}(\hat{S}) + 1 \leq nW(w_1)$. Denote $W = W(w_1)$.

To evaluate the number of operations required by the modified algorithm $\mathcal{A}_4$, we use Theorem 4.1 when considering input and unifying nodes, and results from Sect. 4.3 when considering functional nodes.

To treat all input nodes of the scheme of $\hat{S}$, we do not need any operations. For all at most n unifying nodes with at most n entering edges each, the modified algorithm $\mathcal{A}_4$ makes at most $4n^3W\log_2(n^2W) \leq 8n^3W\log_2(nW)$ comparisons.

The scheme of the circuit $\hat{S}$ contains at most $2n$ functional nodes with at most n entering edges each. To process one functional node, the modified algorithm $\mathcal{A}_4$ makes at most $2n^3W^2$ additions and at most $8n^3W^2\log_2(nW)$ comparisons. In total, the algorithm needs at most $20n^4W^2\log_2(2nW)$ operations to process all functional nodes.

Hence, the modified algorithm $\mathcal{A}_4$ requires at most $28n^4W(w_1)^2\log_2(2nW(w_1))$ operations from B^+ (additions and comparisons), i.e., pseudo-polynomial number of operations (polynomial depending on the dimension n and the numeric parameter $W(w_1)$).

In the software model of computation, for the considered scheme of the circuit and pair of cost functions, the modified algorithm $\mathcal{A}_4$ has pseudo-polynomial time

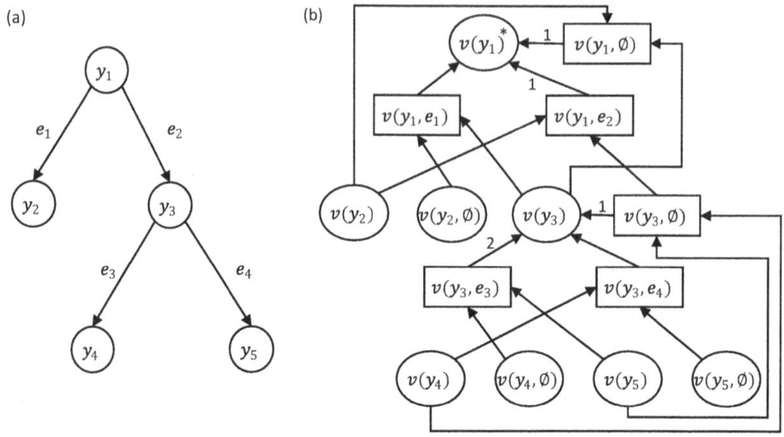

Fig. 18.1 Syntactical circuit $\hat{S}_M$ (**b**) corresponding to tree G (**a**)

complexity $O(n^4W(w_1)^2\log(nW(w_1)))$ (polynomial depending on the length of the problem representation and the numeric parameter $W(w(1))$.

Using Remarks 4.4 and 4.5 we obtain that the number $\max(N_{\Psi^{w_1}}(S), N_{\Psi^{w_2}}(S))$ is an upper bound on the absolute values of the numbers at the inputs of operations from B^+ (additions and comparisons) of the modified algorithm $\mathscr{A}_4$. We know that $\max(N_{\Psi^{w_1}}(S), N_{\Psi^{w_2}}(S)) \le n\max(W(w_1), W(w_2))$. Therefore, in the integer model of computation, for the considered scheme of the circuit and pair of cost functions, the modified algorithm $\mathscr{A}_4$ has pseudo-polynomial time complexity $O(n^4W(w_1)^2\log(nW(w_1))\log(n\max(W(w_1), W(w_2))))$ (polynomial depending on the length of the problem representation and the numeric parameter $W(w_1)$).

18.3 Example

We consider an example (introduced in Part III) of multi-stage and bi-criteria optimization of matchings for the tree G with the root y_1 depicted in Fig. 18.1a, which has five nodes y_1, y_2, y_3, y_4, y_5 and four edges e_1, e_2, e_3, e_4.

We study two weight functions. The first weight function w_1 corresponds the weight 1 to each edge (we maximize the cardinality of matching). The second weight function w_2 corresponds weights $2, 5, 1, 2$ to edges e_1, e_2, e_3, e_4, respectively (we maximize the weight of matching relative to this function).

We construct the syntactical circuit $\hat{S}_M$ (see Fig. 18.1b) and apply to it the procedure of optimization relative to the cost function Ψ^{w_1}. As a result, we have the circuit $\hat{S}_M^{\Psi^{w_1}}$, which can be obtained from the circuit $\hat{S}_M$ by the removal of edges labeled with the number 1 (see Fig. 18.1b). The circuit $\hat{S}_M^{\Psi^{w_1}}$ describes the set containing two for-

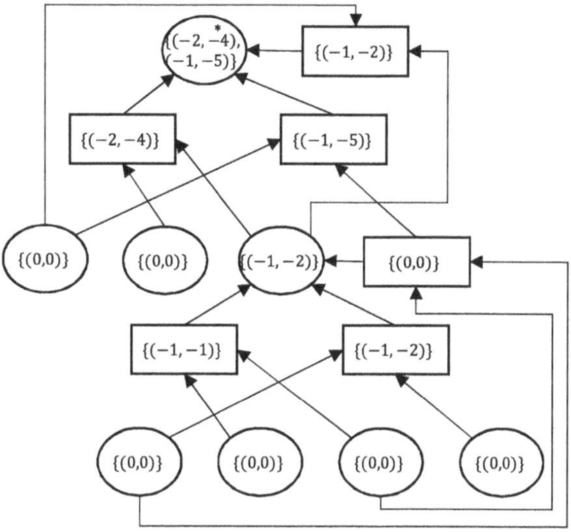

Fig. 18.2 Sets of Pareto optimal points for all nodes of the circuit $\hat{S}_M$

mulas $\hat{\mu}_{y_1,e_1}(\hat{\mu}_{y_3,e_4}(\hat{\lambda}_{y_5,\emptyset}, \hat{\lambda}_{y_4}), \hat{\lambda}_{y_2,\emptyset})$ and $\hat{\mu}_{y_1,e_1}(\hat{\mu}_{y_3,e_3}(\hat{\lambda}_{y_5}, \hat{\lambda}_{y_4,\emptyset}), \hat{\lambda}_{y_2,\emptyset})$ representing matchings $\{e_1, e_4\}$ and $\{e_1, e_3\}$ with maximum cardinality.

We apply to the circuit $\hat{S}_M^{\Psi^{w_1}}$ the procedure of optimization relative to the cost function Ψ^{w_2}. As a result, we have the circuit $\hat{S}_M^{\Psi^{w_1},\Psi^{w_2}}$, which can be obtained from the circuit $\hat{S}_M$ by the removal of edges labeled with the numbers 1 and 2 (see Fig. 18.1b). The circuit $\hat{S}_M^{\Psi^{w_1},\Psi^{w_2}}$ describes the set containing one formula $\hat{\mu}_{y_1,e_1}(\hat{\mu}_{y_3,e_4}(\hat{\lambda}_{y_5,\emptyset}, \hat{\lambda}_{y_4}), \hat{\lambda}_{y_2,\emptyset})$ representing the matching $\{e_1, e_4\}$.

We also apply to the syntactical circuit $\hat{S}_M$ the procedure of bi-criteria optimization relative to the weight functions Ψ^{w_1} and Ψ^{w_2}. As a result, each node of $\hat{S}_M$ is labeled with the set of Pareto optimal points (see Fig. 18.2).

18.4 Experiments

We conduct computational experiments with randomly generated rooted trees, with randomly generated weight function w, and with the weight function c, which assigns the weight 1 to each edge of the tree.

We repeat each experiment 100 times. For each trial, we construct a tree and a weight function w for this tree using the following random procedure. First, we define the range of a possible number of children "# children" and the range of a possible edge weight "weights". Both ranges are given in the form 1-m, where m is a natural number. For the root node, we generate a random number of children in the given range "# children", add edges connecting the root node with the children, and assign weights randomly in the given range "weights" to these edges. Then from all terminal nodes of the constructed tree, we randomly choose one and add children

Table 18.1 Average number of formulas described by the syntactical circuits before optimization ($\emptyset$), after optimization relative to Ψ^c (c), after optimization relative to Ψ^w (w), after optimization relative to Ψ^c and Ψ^w (c, w), and after optimization relative to $\Psi^w$and Ψ^c (w, c) among 100 trials; # children 1–2; weights 1–10

# nodes	Average number of matchings				
	$\emptyset$	c	w	c, w	w, c
10	65.00	4.91	1.31	1.24	1.23
30	5.88×10^{05}	129.13	1.59	1.40	1.47
50	4.58×10^{09}	2.37×10^{03}	3.24	1.70	2.04
70	4.02×10^{13}	2.89×10^{04}	6.34	2.79	3.50
90	4.42×10^{17}	5.05×10^{05}	7.86	3.33	4.13
110	5.15×10^{21}	3.90×10^{07}	11.17	3.73	5.19
130	4.31×10^{25}	5.18×10^{09}	16.90	5.01	8.21
150	2.79×10^{29}	8.11×10^{10}	31.67	7.06	11.29
170	2.91×10^{33}	1.83×10^{12}	48.87	10.80	23.41
190	2.18×10^{37}	5.68×10^{12}	61.85	8.94	16.55
210	1.94×10^{41}	9.00×10^{13}	134.21	12.02	31.52
230	2.06×10^{45}	8.10×10^{14}	177.76	14.57	40.85

to this node, etc. The procedure terminates when the total number of nodes in the constructed tree is equal to the given limit "# nodes". In all random processes, we use uniform distributions.

Tables 18.1 and 18.2 show for # nodes 10, 30, ..., 230, # children 1–2 or 1–15, and weights 1–10 or 1–100, the average number of formulas described by the syntactical circuits before optimization and after optimization relative to all possible sequences of the cost functions Ψ^c and Ψ^w. There were many formulas representing matchings with maximum cardinality and few formulas representing maximum edge weight matchings.

Table 18.3 shows for # nodes 10, 30, ..., 230, # children 1–2 or 1–15, and weights 1–10 or 1–100, the average maximum cardinality of matchings represented by formulas before and after optimization relative to Ψ^w. On average, the maximum cardinality of matchings was greater before optimization than after optimization.

We also considered the bi-criteria optimization of formulas. Table 18.4 shows the minimum, maximum, and the average number of Pareto optimal points for bi-criteria optimization problems relative to Ψ^c and Ψ^w for # nodes 30, 70, ..., 230, # children 1–2, 1–4,..., 1–20, and weights 1–10. On average, trees with # children 1–2 had a larger number of Pareto optimal points. Among all tested scenarios, there were at least one and at most 15 Pareto optimal points. For trees with # children 1–2 and $n = 190, 230$, there were no totally optimal formulas among all experiments—see Table 18.5.

The set of Pareto optimal points for a bi-criteria optimization problem relative to Ψ^c and Ψ^w (# nodes 230, # children 1–2, and weights 1–10) is depicted in Fig. 18.3.

Table 18.2 Average number of formulas described by the syntactical circuits before optimization ($\emptyset$), after optimization relative to Ψ^c (c), after optimization relative to Ψ^w (w), after optimization relative to Ψ^c and Ψ^w (c, w), and after optimization relative to Ψ^w and Ψ^c (w, c) among 100 trials; # children 1–15; weights 1–100

# nodes	$\emptyset$	c	w	c, w	w, c
10	22.00	8.52	1.02	1.01	1.01
30	1.48×10^{04}	699.23	1.11	1.12	1.11
50	1.23×10^{06}	5.91×10^{04}	1.27	1.27	1.27
70	4.41×10^{08}	4.90×10^{06}	1.46	1.45	1.45
90	4.69×10^{11}	4.29×10^{08}	1.35	1.32	1.32
110	2.90×10^{13}	2.86×10^{10}	1.56	1.56	1.56
130	1.14×10^{15}	2.46×10^{12}	1.82	1.86	1.82
150	6.23×10^{17}	2.42×10^{14}	1.85	1.82	1.82
170	9.83×10^{19}	1.92×10^{16}	1.88	1.87	1.85
190	1.46×10^{22}	1.46×10^{18}	2.24	2.22	2.21
210	1.42×10^{24}	1.16×10^{20}	2.56	2.56	2.56
230	1.35×10^{26}	7.38×10^{21}	3.04	3.06	3.04

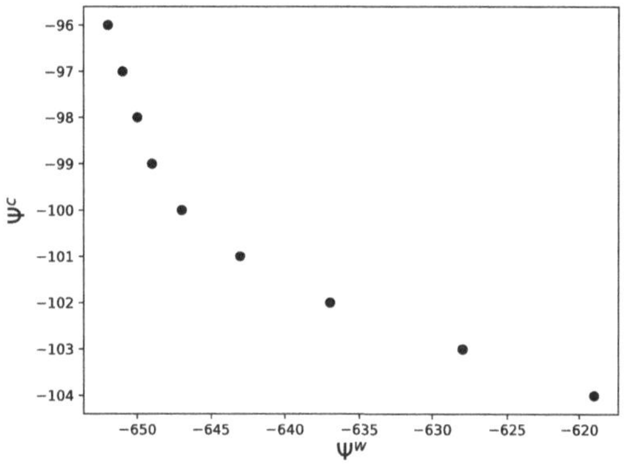

Fig. 18.3 Set of Pareto optimal points for a bi-criteria optimization problem with parameters # nodes 230, # children 1–2, and weights 1–10

18.5 Comparison with Results from Part III

In Part III, for a given rooted tree G with n nodes, we design the forest $D(G)$ with $3n - 1$ nodes and $2n - 1$ edges, which can be constructed in linear time depending on n, and which is used by the algorithms for optimization and counting of matchings.

In this chapter, for the tree G, we design the syntactical circuit $\hat{S}$ with at most $5n$ nodes, which is used by the algorithms for optimization and counting of formulas representing matchings. In fact, we need to construct only the scheme of the circuit $\hat{S}$—see Sect. 2.4. In the general case, there is no linear time algorithm to construct this scheme since it can contain more than a linear number of edges depending on n. Let m be a natural number. We consider a rooted tree with $m^2 + 1$ nodes and

Table 18.3 Average maximum cardinality of matchings represented by formulas before and after optimization relative to Ψ^w among 100 trials

# nodes	# children	Weights	Average maximum matching cardinality	
			before optimization	after optimization
10	1–2	1–10	4.31	4.16
30	1–2	1–10	13.23	12.43
50	1–2	1–10	22.11	20.77
70	1–2	1–10	31.05	29.17
90	1–2	1–10	39.99	37.53
110	1–2	1–10	48.72	46.01
130	1–2	1–10	57.67	54.45
150	1–2	1–10	66.32	62.55
170	1–2	1–10	75.43	71.05
190	1–2	1–10	84.41	79.37
210	1–2	1–10	93.31	88.03
230	1–2	1–10	102.19	96.42
10	1–15	1–100	2.71	2.65
30	1–15	1–100	5.12	5.04
50	1–15	1–100	7.76	7.70
70	1–15	1–100	10.38	10.37
90	1–15	1–100	12.68	12.65
110	1–15	1–100	15.42	15.28
130	1–15	1–100	17.60	17.58
150	1–15	1–100	20.36	20.31
170	1–15	1–100	22.87	22.74
190	1–15	1–100	25.22	25.12
210	1–15	1–100	27.37	27.30
230	1–15	1–100	29.85	29.69

m^2 edges that contains m nodes with m leaving edges each (for $m = 3$, such tree is depicted in Fig. 18.4). The corresponding scheme contains at least m^3 edges.

The algorithm $\mathscr{A}_5$ for counting matchings and the algorithm $\mathscr{A}_6$ for matching optimization considered in Part III require $O(n)$ arithmetical operations each.

The algorithm $\mathscr{A}_2$ for counting formulas and the algorithm $\mathscr{A}_1$ for optimization of formulas (matchings) considered in this chapter require $O(n^2)$ arithmetical operations each. Note that the algorithm $\mathscr{A}_2$ counts not matchings but formulas representing matchings.

Table 18.4 Number of Pareto optimal points (in format $_{min}average_{max}$) for bi-criteria optimization problems relative to Ψ^c and Ψ^w with weights 1–10 among 100 trials

# children	# nodes					
	30	70	110	150	190	230
1–2	$_{1}1.65_{3}$	$_{1}2.95_{7}$	$_{1}3.69_{7}$	$_{1}5.09_{10}$	$_{2}6.05_{11}$	$_{3}7.47_{15}$
1–4	$_{1}1.34_{3}$	$_{1}1.84_{4}$	$_{1}2.03_{4}$	$_{1}2.99_{7}$	$_{1}3.17_{7}$	$_{1}3.82_{8}$
1–6	$_{1}1.21_{3}$	$_{1}1.30_{3}$	$_{1}1.59_{4}$	$_{1}1.76_{4}$	$_{1}2.03_{4}$	$_{1}2.22_{6}$
1–8	$_{1}1.06_{2}$	$_{1}1.13_{3}$	$_{1}1.22_{3}$	$_{1}1.38_{3}$	$_{1}1.43_{4}$	$_{1}1.38_{3}$
1–10	$_{1}1.02_{2}$	$_{1}1.09_{2}$	$_{1}1.13_{3}$	$_{1}1.17_{3}$	$_{1}1.26_{3}$	$_{1}1.20_{3}$
1–12	$_{1}1.02_{2}$	$_{1}1.08_{2}$	$_{1}1.08_{2}$	$_{1}1.05_{2}$	$_{1}1.11_{3}$	$_{1}1.13_{2}$
1–14	$_{1}1.03_{2}$	$_{1}1.03_{2}$	$_{1}1.02_{2}$	$_{1}1.06_{2}$	$_{1}1.08_{2}$	$_{1}1.05_{2}$
1–16	$_{1}1.00_{1}$	$_{1}1.02_{2}$	$_{1}1.01_{2}$	$_{1}1.06_{2}$	$_{1}1.04_{2}$	$_{1}1.06_{2}$
1–18	$_{1}1.00_{1}$	$_{1}1.01_{2}$	$_{1}1.04_{2}$	$_{1}1.01_{2}$	$_{1}1.04_{2}$	$_{1}1.03_{2}$
1–20	$_{1}1.02_{2}$	$_{1}1.01_{2}$	$_{1}1.03_{2}$	$_{1}1.03_{3}$	$_{1}1.03_{3}$	$_{1}1.04_{2}$

Table 18.5 Number of trees with totally optimal formulas (matchings) relative to c and w with weights 1–10 among 100 trials

# children	# nodes					
	30	70	110	150	190	230
1–2	45	8	5	1	0	0
1–4	72	37	29	9	10	6
1–6	81	74	52	47	33	32
1–8	94	88	79	67	65	66
1–10	98	91	88	84	77	83
1–12	98	92	92	95	90	87
1–14	97	97	98	94	92	95
1–16	100	98	99	94	96	94
1–18	100	99	96	99	96	97
1–20	98	99	97	98	98	96

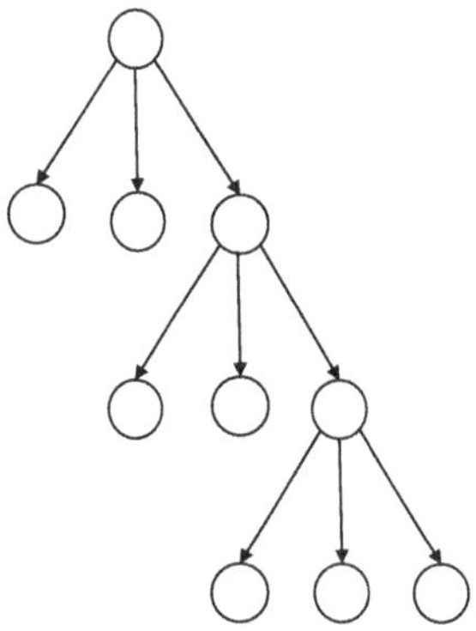

Fig. 18.4 Rooted tree with $m^2 + 1$ nodes and m^2 edges, where $m = 3$

The algorithm $\mathscr{A}_8$ for the construction of POPs for matchings considered in Part III requires $O(n^6 W^2 \log(nW))$ arithmetical operations, where

$$W = W(w_1) = \max\{|w_1(e)| : e \in E(G)\} .$$

The modified algorithm $\mathscr{A}_4$ for the construction of POPs for formulas (matchings) considered in this chapter requires $O(n^4 W^2 \log(nW))$ arithmetical operations.

The number of operations for the algorithm $\mathscr{A}_8$ is overestimated. In particular, to evaluate the number of operations, we used a bound from [1] (see Proposition 16.2) instead of the improved bounds considered in Sect. 4.3. One can obtain for the algorithm $\mathscr{A}_8$ an upper bound on the number of operations similar to the upper bound for the modified algorithm $\mathscr{A}_4$ obtained in this chapter.

As a result, we see that the direct study of combinatorial optimization problems (as in Part III) can lead to algorithms, which time complexity is similar or better than the time complexity of algorithms based on the circuit approach (as in this chapter). However, the direct study requires additional effort to design and analyze algorithms for optimization and counting.

The experiments considered in this chapter were done with different randomly generated trees and weight functions than the experiments in Part III. However, the results of experiments obtained in this chapter and in Part III are similar.

References

1. AbouEisha, H., Amin, T., Chikalov, I., Hussain, S., Moshkov, M.: Extensions of Dynamic Programming for Combinatorial Optimization and Data Mining, Intelligent Systems Reference Library, vol. 146. Springer (2019)
2. Mankowski, M., Moshkov, M.: Bi-criteria optimization of matchings in trees with application to kidney exchange. In: 26th International Workshop on Concurrency, Specification and Programming, CS&P 2017, Warsaw, Poland, September 25–27, 2017 (2017). http://csp2017.mimuw.edu.pl/data/uploads/papers/CSP2017_paper_1.pdf
3. Mankowski, M., Moshkov, M.: Multi-stage optimization of matchings in trees with application to kidney exchange. In: Polkowski, L., Yao, Y., Artiemjew, P., Ciucci, D., Liu, D., Slezak, D., Zielosko, B. (eds.) Proceedings of the International Joint Conference on Rough Sets (IJCRS 2017), Part I, Olsztyn, Poland, July 3–7, 2017. Lecture Notes in Computer Science, vol. 10313, pp. 123–130. Springer (2017)

Chapter 19
0/1 Knapsack Problem

We consider the 0/1 knapsack problem consisting of capacity limit and set of items with associated weights and values. None of the items can be duplicated. The task is to find a set of items with the maximum total value in which total weight does not exceed the capacity limit. This problem is NP-hard [1] and has pseudo-polynomial dynamic programming solution [2]. We also associate the environmental cost to each of the items with the aim of minimizing the total cost.

In this chapter, we design conventional and syntactical circuits for the 0/1 knapsack problem, describe two cost functions for the syntactical circuit, evaluate the number of operations and the time required by the algorithms for optimization and counting, discuss an example, and show experimental results.

19.1 Definition of Circuit and Cost Functions

Let us consider n, $n \geq 1$, items $e_1, e_2, \ldots, e_n$, which have positive integer weights $w_1, w_2, \ldots, w_n$. Also, there is a positive integer W that is the maximum capacity. We now describe the set $M(n, W)$ of all sets of items such that their total weight is at most W. For $0 \leq i \leq n$ and $0 \leq w \leq W$, we denote by $M(i, w)$ the set of all sets of items from $\{e_1, e_2, \ldots, e_i\}$ whose total weight is at most w. Then $M(0, w) = \{\emptyset\}$ for $w = 0, 1, \ldots, W$, and $M(i, 0) = \{\emptyset\}$ for $i = 0, 1, \ldots, n$. For $1 \leq i \leq n$ and $1 \leq w \leq W$, we have:

$$M(i, w) = \begin{cases} M(i-1, w), & \text{if } w_i > w\,, \\ M(i-1, w) \cup (M(i-1, w-w_i) \otimes \{e_i\}), & \text{if } w_i \leq w\,, \end{cases}$$

where, for any nonempty set of sets of items A, $A \otimes \{e_i\} = \{\alpha \cup \{e_i\} : \alpha \in A\}$.

M. Mankowski and M. Moshkov, *Dynamic Programming Multi-Objective Combinatorial Optimization*, Studies in Systems, Decision and Control 331,
https://doi.org/10.1007/978-3-030-63920-4_19

The number n is the dimension of the problems under consideration related to the optimization of sets of items. To define a cost function (total value or total environmental cost), we need to specify n nonnegative real numbers (values or environmental costs) corresponding to items. To describe an instance of the problem, we need the capacity limit W, n weights, and one or two cost functions. Note that n is at most the number of numeric parameters of the problem.

Let us assume that values and environmental costs are nonnegative integers. Then the binary representations of the capacity limit W, n weights, and n or $2n$ numbers defining one or two cost functions form the representation of the problem instance (we will omit the word instance). It is clear that n is at most the length of the problem representation.

Set $U = \bigcup_{0 \le i \le n, \, 0 \le w \le W} M(i, w)$. We now describe a circuit S over the set U, which represents the set $M(n, W)$.

For $0 \le i \le n$, the circuit S contains the input node $v(i, 0)$ labeled with $\{\lambda_{i,0}\}$, where $\lambda_{i,0} = \emptyset$. For $0 \le w \le W$, the circuit S contains the input node $v(0, w)$ labeled with $\{\lambda_{0,w}\}$, where $\lambda_{0,w} = \emptyset$.

For $1 \le i \le n$ and $1 \le w \le W$, the circuit S contains a unifying node $v(i, w)$ labeled with $\cup$, a functional node $v(i, w, 0)$ labeled with the function $\mu_{i,w,0}(x) = x$, and, if $w_i \le w$, the circuit S also contains a functional node $v(i, w, i)$ labeled with the function $\mu_{i,w,i}(x) = x \cup \{e_i\}$. The node $v(n, W)$ is considered as the output and is also labeled with $*$.

For $1 \le i \le n$ and $1 \le w \le W$, each unifying node $v(i, w)$ has at most two entering edges, always from $v(i, w, 0)$ and from $v(i, w, i)$ if $w_i \le w$. For $1 \le i \le n$ and $1 \le w \le W$, the functional node $v(i, w, 0)$ has one entering edge from the node $v(i-1, w)$ and, if $w_i \le w$, then the functional node $v(i, w, i)$ has one entering edge from the node $v(i-1, w-w_i)$. We will not label these edges with the number 1.

One can show that, for any input or unifying node $v(i, w)$, $S(v(i, w)) = M(i, w)$. For $v(i, w, i) \in V_f(S)$, $S(v(i, w, i)) = M(i-1, w-w_i) \otimes \{e_i\}$ and, for $v(i, w, 0) \in V_f(S)$, $S(v(i, w, 0)) = M(i-1, w)$.

The circuit S contains $n + W + 1$ input nodes, at most nW unifying nodes with at most two entering edges each, and at most $2nW$ functional nodes with one entering edge each. Hence, the circuit S contains at most $3nW + n + W + 1$ nodes and at most $4nW$ edges.

The circuit S is not a circuit without repetitions: both input and functional nodes might share some of the elements. We correspond to S the syntactical circuit $\hat{S}'$, which is a circuit without repetitions (see Sect. 2.3).

We denote by $\hat{S}$ the circuit obtained from $\hat{S}'$ by the removal of all nodes such that there is no directed path from the considered node to the output. We also remove all edges that are incident to the removed nodes. It is clear that $\hat{S}(v(n, W)) = \hat{S}'(v(n, W))$.

In the syntactical circuit $\hat{S}$, we use element symbols $\hat{\lambda}_{0,w}$ and $\hat{\lambda}_{i,0}$ instead of elements $\lambda_{0,w}$ and $\lambda_{i,0}$, and functional symbols (and functions) $\hat{\mu}_{i,w,0}$ and $\hat{\mu}_{i,w,i}$ instead of functions $\mu_{i,w,0}$ and $\mu_{i,w,i}$.

We now define a cost function $\Psi^{(t)}$, $t = 1, 2$, for the circuit $\hat{S}$ given by a function $\psi^{(t)}$ and function $\psi_v^{(t)}$ for each $v \in V_f(\hat{S})$. Each item $e_i \in \{e_1, \ldots, e_n\}$ has corresponding value d_i and environmental cost c_i that are nonnegative real numbers.

We consider the problem of maximization of the total value. The cost function $\Psi^{(1)}$ is defined by the functions $\psi^{(1)}(\hat{\lambda}_{i,0}) = \psi^{(1)}(\hat{\lambda}_{0,w}) = 0$, $0 \le i \le n$, $0 \le w \le W$, $\psi^{(1)}_{v(i,w,0)}(x) = x$, $v(i, w, 0) \in V_f(\hat{S})$, and $\psi^{(1)}_{v(i,w,i)}(x) = x - d_i$, $v(i, w, i) \in V_f(\hat{S})$.

To compute $\psi^{(1)}$, we do not need any operations, while to compute $\psi_v^{(1)}$, $v \in V_f(\hat{S})$, we need at most one operation of subtraction. The function $\Psi^{(1)}$ is a strictly increasing cost function and, if all values $d_1, \ldots, d_n$ are nonnegative integers, then $\Psi^{(1)}$ is a nonpositive integer cost function for $\hat{S}$.

We also consider the problem of minimization of the total environmental cost. The cost function $\Psi^{(2)}$ is defined by the functions $\psi^{(2)}(\hat{\lambda}_{i,0}) = \psi^{(2)}(\hat{\lambda}_{0,w}) = 0$, $0 \le i \le n$, $0 \le w \le W$, $\psi^{(2)}_{v(i,w,0)}(x) = x$, $v(i, w, 0) \in V_f(\hat{S})$, and $\psi^{(2)}_{v(i,w,i)}(x) = x + c_i$, $v(i, w, i) \in V_f(\hat{S})$.

To compute $\psi^{(2)}$, we do not need any operations, while to compute $\psi_v^{(2)}$, $v \in V_f(\hat{S})$, we need at most one operation of addition. The function $\Psi^{(2)}$ is a strictly increasing cost function and, if all costs $c_1, \ldots, c_n$ are nonnegative integers, then $\Psi^{(2)}$ is a nonnegative integer cost function for $\hat{S}$.

The considered cost functions are adapted (see Sect. 2.4). They are either equal to constants or variables or described by formulas over the set of elementary operations $B = \{x + y, x - y\}$.

Let δ be a formula from $\hat{S}(v^*)$. Then the element $a(\delta)$ from $S(v^*)$ represented by δ is a subset of the set $\{e_1, \ldots, e_n\}$. It is easy to see that $\Psi^{(1)}(\delta) = -\Sigma_{e_i \in a(\delta)} d_i$ and $\Psi^{(2)}(\delta) = \Sigma_{e_i \in a(\delta)} c_i$. Therefore, $\Psi^{(1)}$ and $\Psi^{(2)}$ are proper cost functions for the circuit $\hat{S}$ relative to the circuit S.

19.2 Time Complexity of Algorithms

The algorithms $\mathscr{A}_1$, $\mathscr{A}_2$, and $\mathscr{A}_4$ work with the scheme of the circuit $\hat{S}$ (see Sect. 2.4). For $1 \le i \le n$ and $1 \le w \le W$, we need to check if $w_i \le w$ to recognize if the functional node $v(i, w, i)$ belongs to the circuit $\hat{S}'$. The maximum absolute value of the integers at the inputs of the comparisons is equal to $q = \max\{w_1, \ldots, w_n, W\}$. The time complexity of these operations is $O(nW)$ in the software model of computation and $O(nW \log q)$ in the integer model of computation.

After nW comparisons, the circuit $\hat{S}'$ will be completely described and we will be able to construct the scheme of $\hat{S}'$ in a linear time depending on the number of its nodes and edges, i.e., in $O(nW)$. Using breadth-first search from the node $v(n, W)$ in the scheme of $\hat{S}'$ with the reverse direction of the edges, we can find in the scheme of $\hat{S}'$ all nodes such that there is no directed path from the considered node to the output and construct the scheme of $\hat{S}$ in $O(nW)$ time. In total, we need $O(nW)$ time to construct the scheme of $\hat{S}$.

Note that the algorithms $\mathscr{A}_1$ and $\mathscr{A}_2$ can also work with the schemes of edge-preserving subcircuits of $\hat{S}$ obtained as a result of the algorithm $\mathscr{A}_1$ work—see Remark 3.1.

Denote $B^+ = B \cup \{x : y\}$, where $x : y$ is the operation of comparison with values $x < y$, $x = y$, and $x > y$.

Let us assume that d_i and c_i, $i = 1, \ldots, n$, are nonnegative integers. Denote $\rho_1 = \max\{d_1, \ldots, d_n\}$ and $\rho_2 = \max\{c_1, \ldots, c_n\}$. Let $t \in \{1, 2\}$. One can show that $N_{\Psi^{(t)}}(\hat{S}) \leq \rho_t n$, where $N_{\Psi^{(t)}}(\hat{S}) = \max\{|\Psi^{(t)}(\delta)| : \delta \in U(\hat{S})\}$ and $U(\hat{S}) = \bigcup_{v \in V(\hat{S})} \hat{S}(v)$. One can also show that $\log_2 \rho_t$ is at most the length of the problem representation if the problem uses the cost function $\Psi^{(t)}$.

We can use the algorithm $\mathscr{A}_1$ for multi-stage optimization of formulas (sets of items). Using Theorem 3.1, we evaluate the number of operations required by the algorithm $\mathscr{A}_1$ working with the scheme of an edge-preserving subcircuit T of the circuit $\hat{S}$ and a cost function $\Psi^{(t)}$ for $\hat{S}$, where $t \in \{1, 2\}$.

The scheme of T contains at most $n + W + 1$ input nodes. We do not need any operations to process these nodes.

The scheme of T contains at most nW unifying nodes with at most two entering edges each. To treat these nodes, the algorithm $\mathscr{A}_1$ makes at most $(4 - 1)nW \leq 3nW$ comparisons.

The scheme of T contains at most $2nW$ functional nodes. For each functional node v, the algorithm makes at most one operation (addition or subtraction) to compute $\psi_v^{(t)}$. The total number of operations for functional nodes is at most $2nW$. The overall number of operations from B^+ (additions, subtractions, and comparisons) made by $\mathscr{A}_1$ is at most $5nW$, i.e., it is pseudo-polynomial (polynomial depending on the dimension n and the numeric parameter W).

In the software model of computation, for the considered scheme of the circuit and cost function, the algorithm $\mathscr{A}_1$ has pseudo-polynomial time complexity $O(nW)$ (polynomial depending on the dimension n and the numeric parameter W).

Using Remark 3.2 we obtain that $\rho_t n$ is an upper bound on the absolute values of the numbers at the inputs of operations of the algorithm $\mathscr{A}_1$. The algorithm uses only the operations of addition, subtraction, and comparison. Therefore, in the integer model of computation, for the considered scheme of the circuit and the cost function, the algorithm $\mathscr{A}_1$ has pseudo-polynomial time complexity $O(nW \log(\rho_t n))$ (polynomial depending on the length of the problem representation and the numeric parameter W).

To count all formulas described by circuit T, we apply the algorithm $\mathscr{A}_2$. We use Theorem 3.2 to evaluate the number of operations required by $\mathscr{A}_2$. The scheme of T contains at most nW unifying nodes with at most two entering edges each. Therefore, we need at most nW additions to treat these unifying nodes. For each of at most $2nW$ functional nodes, we do not need any operations. To treat the input nodes, we do not need any operations. Thus, the total number of operations from B^+ made by $\mathscr{A}_2$ is at most nW, i.e., it is pseudo-polynomial (polynomial depending on the dimension n and the numeric parameter W).

In the software model of computation, for the considered scheme of the circuit, the algorithm $\mathcal{A}_2$ has pseudo-polynomial time complexity $O(nW)$ (polynomial depending on the dimension n and the numeric parameter W).

We will not consider the time complexity of the algorithm $\mathcal{A}_2$ in the integer model of computation since we cannot find reasonable upper bounds on the number of formulas described by the circuit $\hat{S}$.

Let us assume that d_i and c_i, $i = 1, \ldots, n$, are nonnegative integers. We consider two distinct cost functions $\Theta^{(1)}, \Theta^{(2)} \in \{\Psi^{(t)} : t = 1, 2\}$.

We can use the algorithm $\mathcal{A}_4$ to find the set of Pareto optimal points for the problem of bi-criteria optimization of formulas (sets of items) relative to $\Theta^{(1)}$ and $\Theta^{(2)}$. We denote $M = M_{\Theta^{(1)},\Theta^{(2)}}(\hat{S}) = \max\{|Par(t_{\Theta^{(1)},\Theta^{(2)}}(\hat{S}, v))| : v \in V(\hat{S})\}$.

To evaluate the number of operations of the algorithm $\mathcal{A}_4$, we use Theorem 4.1. To treat the input nodes of the scheme of $\hat{S}$, we do not need any operations.

The scheme of $\hat{S}$ contains at most $2nW$ functional nodes with one entering edge. To process one functional node, the algorithm $\mathcal{A}_4$ makes at most $2M$ operations from B and at most $4M \log_2 M$ comparisons. Therefore, the algorithm needs at most $12nWM \log_2(2M)$ operations from B^+ to process all functional nodes.

For all at most nW unifying nodes with at most two entering edges each, the algorithm $\mathcal{A}_4$ makes at most $8nWM \log_2(2M)$ comparisons.

Hence, the algorithm $\mathcal{A}_4$ requires at most $20nWM \log_2(2M)$ operations from B^+ (additions, subtractions, and comparisons).

Let $\Theta^{(1)} = \Psi^{(t)}$, $t \in \{1, 2\}$. By Proposition 4.4, $M \leq \rho_t n + 1$. Then, the algorithm $\mathcal{A}_4$ requires at most $20nW(\rho_t n + 1) \log_2(2\rho_t n + 2)$ operations from B^+, i.e., pseudo-polynomial number of operations (polynomial depending on the dimension n and the numeric parameters W and ρ_t).

In the software model of computation, for the considered scheme of the circuit and pair of cost functions, the algorithm $\mathcal{A}_4$ has pseudo-polynomial time complexity $O(n^2 W\rho_t \log(\rho_t n))$ (polynomial depending on the length of the problem representation and the numeric parameters W and ρ_t).

Using Remark 4.4 we obtain that the number $\max(N_{\Psi^{(1)}}(\hat{S}), N_{\Psi^{(2)}}(\hat{S}))$ is an upper bound on the absolute values of the numbers at the inputs of operations of the algorithm $\mathcal{A}_4$. This algorithm uses only operations of addition, subtraction, and comparison. We know that $\max(N_{\Psi^{(1)}}(\hat{S}), N_{\Psi^{(2)}}(\hat{S})) \leq \max(\rho_1, \rho_2)n$. Therefore, in the integer model of computation, for the considered scheme of the circuit and pair of cost functions, the algorithm $\mathcal{A}_4$ has pseudo-polynomial time complexity $O(n^2 W\rho_t \log(\rho_t n) \log(\max(\rho_1, \rho_2)n))$ (polynomial depending on the length of problem representation and the numeric parameters W and ρ_t).

19.3 Example

We consider three items e_1, e_2, e_3, with corresponding weights $w_1 = 4, w_2 = 1$, $w_3 = 5$, values $d_1 = 1, d_2 = 0, d_3 = 3$, and environmental costs $c_1 = 3, c_2, = 5$, $c_3 = 2$. The capacity limit W is equal to 5.

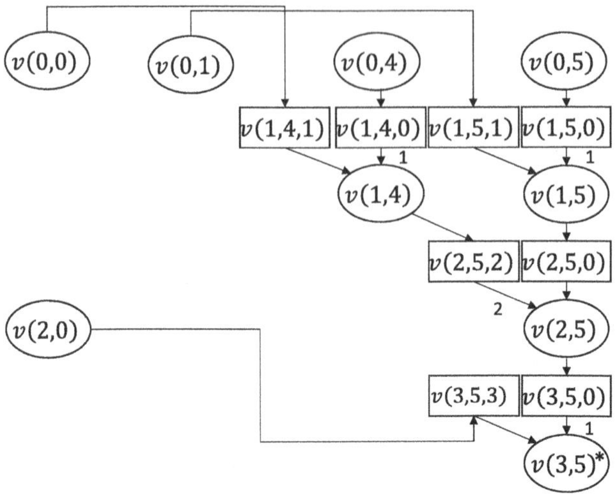

Fig. 19.1 Syntactical circuit $\hat{S}_{KP}$

We construct the circuit S_{KP}, which describes all sets of items for the considered problem. This circuit is a circuit with repetitions. We transform it into the syntactical circuit $\hat{S}_{KP}$ (see Fig. 19.1). We apply to the circuit $\hat{S}_{KP}$ the procedure of optimization relative to $\Psi^{(1)}$. As a result, we have the circuit $\hat{S}_{KP}^{\Psi^{(1)}}$, which can be obtained from the circuit $\hat{S}_{KP}$ by the removal of edges labeled with the number 1. The circuit $\hat{S}_{KP}^{\Psi^{(1)}}$ describes the set containing only one formula $\delta = \hat{\mu}_{3,5,3}(\hat{\lambda}_{2,0})$ for which $a(\delta) = \{e_3\}$. We apply to the circuit $\hat{S}_{KP}^{\Psi^{(1)}}$ the procedure of optimization relative to $\Psi^{(2)}$. As a result, we obtain the circuit $\hat{S}_{KP}^{\Psi^{(1)},\Psi^{(2)}}$, which can be obtained from the circuit $\hat{S}_{KP}$ by the removal of edges labeled with the numbers 1 and 2. The circuit $\hat{S}_{KP}^{\Psi^{(1)},\Psi^{(2)}}$ describes the same set of formulas as the circuit $\hat{S}_{KP}^{\Psi^{(1)}}$.

We also apply the procedure of bi-criteria optimization relative to $\Psi^{(1)}$ and $\Psi^{(2)}$ to the circuit $\hat{S}_{KP}$. The set of Pareto optimal points for each node of $\hat{S}_{KP}$ is depicted in Fig. 19.2.

19.4 Experiments

We experimentally tested various scenarios for optimization of formulas representing sets of items. Each experiment was repeated 10 times.

We performed experiments with different numbers of items n and with different capacity limits W. For $i = 1, \ldots, n$, we generated randomly with the uniform distribution a positive integer weight w_i from the range $[1, w_{max}]$, a nonnegative integer value d_i from the range $[0, d_{max}]$, and a nonnegative integer environmental cost c_i from the range $[0, c_{max}]$.

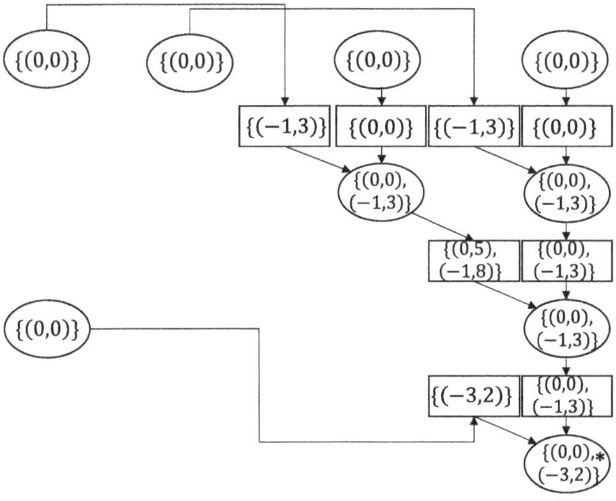

Fig. 19.2 Sets of Pareto optimal points for all nodes of circuit $\hat{S}_{KP}$

Table 19.1 shows for $n = 50, 100, \ldots, 300$, $W = 10, 100$, $w_{max} = 10$, $d_{max} = 100$, and $c_{max} = 100$, the average number of formulas described by the syntactical circuits before optimization, and after optimization relative to all possible sequences of the cost functions $\Psi^{(1)}$ and $\Psi^{(2)}$. Increasing the capacity limit W from 10 to 100 increased the average number of formulas described by the circuit before optimization. For both limits $W = 10, 100$, there were few optimal formulas relative to $\Psi^{(1)}$ or $\Psi^{(2)}$.

The number of formulas can be greater than the number of sets of items. However, further results obtained with the help of the algorithms $\mathscr{A}_1$ and $\mathscr{A}_4$ are true both for formulas and sets of items.

Table 19.2 shows the average value of $\Psi^{(1)}$ after optimization relative to $\Psi^{(1)}$ and relative to $\Psi^{(2)}$, $\Psi^{(1)}$, and the average value of $\Psi^{(2)}$ after optimization relative to $\Psi^{(2)}$ and relative to $\Psi^{(1)}$, $\Psi^{(2)}$. Regardless the capacity limit W, the minimum value of $\Psi^{(2)}$ was 0 in all tested cases after the optimization relative to $\Psi^{(2)}$. This minimum is reachable by formulas describing the empty set of items.

We also considered bi-criteria optimization of formulas (sets of items). Table 19.3 shows the minimum, maximum, and the average number of Pareto optimal points for bi-criteria optimization problems relative to $\Psi^{(1)}$ and $\Psi^{(2)}$ for $n = 50, 100, \ldots, 300$, $W = 100, 150, 200$, $w_{max} = 10$, $d_{max} = 100$, and $c_{max} = 100$. With the increase of n and W, the average number of Pareto optimal points increased. Among all tested scenarios, there were at least 114 and at most 1490 Pareto optimal points. There were no totally optimal formulas (sets of items) among all experiments—see Table 19.4. The set of Pareto optimal points for a bi-criteria optimization of formulas relative to $\Psi^{(1)}$ and $\Psi^{(2)}$ ($n = 300$, $W = 200$, $w_{max} = 10$, $d_{max} = 100$, and $c_{max} = 100$) is depicted in Fig. 19.3.

Table 19.1 Average number of formulas described by syntactical circuits before optimization and after optimization relative to all possible sequences of cost functions $\Psi^{(1)}$ and $\Psi^{(2)}$ with $W = 10, 100$, $w_{max} = 10$, $d_{max} = 100$, and $c_{max} = 100$ among 10 trials

$W = 10$					
n	$\emptyset$	$\Psi^{(1)}$	$\Psi^{(2)}$	$\Psi^{(1)}, \Psi^{(2)}$	$\Psi^{(2)}, \Psi^{(1)}$
50	1.30×10^{04}	1.30	1.60	1.00	1.00
100	3.54×10^{05}	1.00	2.10	1.00	1.00
150	3.26×10^{06}	1.00	3.90	1.00	1.00
200	1.23×10^{07}	1.00	3.50	1.00	1.00
250	5.28×10^{07}	1.20	4.90	1.00	1.00
300	3.10×10^{08}	1.20	6.20	1.00	1.10
$W = 100$					
n	$\emptyset$	$\Psi^{(1)}$	$\Psi^{(2)}$	$\Psi^{(1)}, \Psi^{(2)}$	$\Psi^{(2)}, \Psi^{(1)}$
50	7.18×10^{13}	1.30	1.60	1.00	1.00
100	3.28×10^{21}	1.10	2.20	1.00	1.00
150	1.17×10^{26}	1.50	7.20	1.00	1.00
200	3.72×10^{29}	1.50	6.10	1.00	1.00
250	5.05×10^{32}	1.40	6.60	1.00	1.00
300	7.05×10^{35}	1.40	13.00	1.00	1.10

Table 19.2 Average value of $\Psi^{(1)}$ after optimization relative to $\Psi^{(1)}$ and relative to $\Psi^{(2)}, \Psi^{(1)}$, and average value of $\Psi^{(2)}$ after optimization relative to $\Psi^{(2)}$ and relative to $\Psi^{(1)}, \Psi^{(2)}$ with $W = 10, 100$, $w_{max} = 10$, $d_{max} = 100$, and $c_{max} = 100$ among 10 trials

$W = 10$				
n	$\Psi^{(1)}$	$\Psi^{(2)}, \Psi^{(1)}$	$\Psi^{(2)}$	$\Psi^{(1)}, \Psi^{(2)}$
50	−407.70	−28.10	0	245.50
100	−583.60	−36.80	0	380.00
150	−668.40	−65.10	0	465.30
200	−741.80	−72.10	0	488.30
250	−791.30	−76.80	0	563.80
300	−821.50	−89.30	0	433.90
$W = 100$				
n	$\Psi^{(1)}$	$\Psi^{(2)}, \Psi^{(1)}$	$\Psi^{(2)}$	$\Psi^{(1)}, \Psi^{(2)}$
50	-1.64×10^{03}	−28.10	0	1.17×10^{03}
100	-2.31×10^{03}	−38.70	0	1.74×10^{03}
150	-2.81×10^{03}	−100.30	0	2.08×10^{03}
200	-3.20×10^{03}	−94.30	0	2.30×10^{03}
250	-3.46×10^{03}	−93.50	0	2.47×10^{03}
300	-3.76×10^{03}	−157.50	0	2.60×10^{03}

Table 19.3 Number of Pareto optimal points (in format ${}_{min}average_{max}$) among 10 trials for bi-criteria optimization problems relative to $\Psi^{(1)}$, $\Psi^{(2)}$

n	W		
	100	150	200
50	${}_{114}211.70_{270}$	${}_{235}313.40_{402}$	${}_{279}371.40_{515}$
100	${}_{279}349.70_{488}$	${}_{384}510.80_{666}$	${}_{470}566.80_{728}$
150	${}_{435}543.60_{612}$	${}_{594}697.90_{832}$	${}_{746}870.80_{1005}$
200	${}_{372}547.80_{719}$	${}_{605}838.70_{1100}$	${}_{846}980.10_{1085}$
250	${}_{441}631.30_{735}$	${}_{712}962.30_{1201}$	${}_{984}1269.70_{1470}$
300	${}_{586}778.40_{975}$	${}_{789}1077.60_{1443}$	${}_{1238}1361.80_{1490}$

Table 19.4 Number of cases with totally optimal formulas relative to $\Psi^{(1)}$ and $\Psi^{(2)}$ among 10 trials

n	W		
	100	150	200
50	0	0	0
100	0	0	0
150	0	0	0
200	0	0	0
250	0	0	0
300	0	0	0

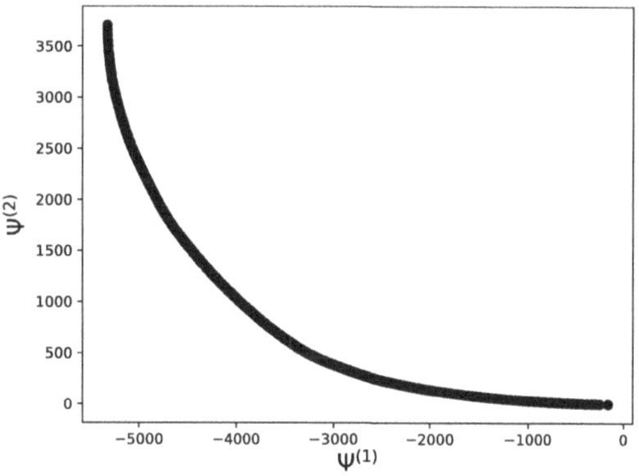

Fig. 19.3 Set of Pareto optimal points for a bi-criteria optimization of formulas relative to $\Psi^{(1)}$ and $\Psi^{(2)}$ ($n = 300$, $W = 200$, $w_{max} = 10$, $d_{max} = 100$, and $c_{max} = 100$)

References

1. Garey, M.R., Johnson, D.S.: Computers and Intractability; A Guide to the Theory of NP-Completeness. W. H. Freeman & Co. (1990)
2. Kellerer, H., Pferschy, U., Pisinger, D.: Knapsack Problems. Springer (2004)

Final Remarks

This book introduced a fairly universal approach to the design and analysis of optimization algorithms for multi-objective combinatorial optimization problems. We proposed the circuits without repetitions along with the increasing and strictly increasing cost functions as a model for such problems. We designed the algorithms for multi-stage and bi-criteria optimization and for counting the solutions in the framework of this model.

We studied nine known combinatorial optimization problems using conventional circuits without repetitions. We separately investigated the problem of matching optimization in a tree for which we did not find an appropriate conventional circuit without repetitions. For this problem, we designed algorithms for multi-stage and bi-criteria optimization and for counting the solutions from scratch. We also studied two combinatorial optimization problems (optimization of matchings in trees and 0/1 knapsack problem) using syntactical circuits without repetitions.

Future studies will be devoted to the consideration of new combinatorial optimization problems and to the design of algorithms for the construction of Pareto optimal elements corresponding to Pareto optimal points.

M. Mankowski and M. Moshkov, *Dynamic Programming Multi-Objective Combinatorial Optimization*, Studies in Systems, Decision and Control 331,
https://doi.org/10.1007/978-3-030-63920-4

Index

M. Mankowski and M. Moshkov, *Dynamic Programming Multi-Objective Combinatorial Optimization*, Studies in Systems, Decision and Control 331,
https://doi.org/10.1007/978-3-030-63920-4

The manufacturer's authorised representative in the EU is Springer Nature Customer Service Centre GmbH, Europaplatz 3, 69115 Heidelberg, Germany. If you have any concerns regarding our products, please contact ProductSafety@springernature.com

Printed and bound by CPI Group (UK) Ltd, Croydon, CR0 4YY
30/09/2026
02225706-0002